The Railway Goods Shed
Warehouse in England

T.W. PARKER & Co

The Railway Goods Shed and Warehouse in England

John Minnis with Simon Hickman

Front cover
Langwathby, one of the standardised designs erected by the MR on the Settle & Carlisle line in 1876.
[DP169031]

Inside front cover
No 4 Bonded Stores and Grain Warehouse, Burton Upon Trent, Staffordshire.
[DP172885]

Frontispiece
Forth Goods station, NER, photographed on 24 July 1948. Newcastle Central passenger station is to the left. Forth Goods was built in 1870 and extended to the east in 1891–4. All that is left of it today are the goods offices, shared with the District Engineer and built in 1903, which stand behind the main goods shed, and the 1906 concrete-framed extension to the shed itself, to the right of the office block.
[EAW017454]

Opposite Foreword
Larger goods sheds often had substantial cellars for warehousing. This view is looking along the length of the cellars at Worcester's MR goods shed and highlights how strong such structures can be in spatial terms.
[DP173449]

Inside back cover
Goods shed at Harrow Road, Great Central Railway, from The Railway Engineer, *July 1906.*

Published by Historic England, The Engine House, Fire Fly Avenue, Swindon SN2 2EH
www.HistoricEngland.org.uk

Historic England is a Government service championing England's heritage and giving expert, constructive advice, and the English Heritage Trust is a charity caring for the National Heritage Collection of more than 400 historic properties and their collections.

First published 2016

ISBN 978-1-84802-328-4 (paperback)
ISBN 978-1-84802-329-1 (e-book), Version 1.0

British Library Cataloguing in Publication data
A CIP catalogue record for this book is available from the British Library.

For more information about images from the Archive, contact Archives Services Team, Historic England, The Engine House, Fire Fly Avenue, Swindon SN2 2EH; telephone (01793) 414600.

Brought to publication by Sarah Enticknap, Publishing, Historic England.

Typeset in Georgia Pro Light 9.25/13pt

Edited by John King
Page layout by Pauline Hull

Printed in Belgium by Graphius – Deckers Snoeck

Contents

Foreword vii
Preface viii
Acknowledgements x
Abbreviations xi

1 How a goods shed functioned 1

2 The origins and evolution of the goods shed 9

3 Plan forms 25

4 Company designs 33

5 Large goods sheds and warehouses 59

6 The 20th-century goods shed and warehouse 79

7 Conservation *Simon Hickman* 91

Gazetteer 100
Notes 125
Bibliography 126

Foreword

Railway goods sheds and warehouses are building types that are no longer used for their original purpose, as technological change, such as the move to containers, has removed their role in handling Britain's freight traffic. Their obsolescence does not, however, render them insignificant. For 130 years, these buildings played a vital role in the British economy. In the major cities and towns of the British Isles, they were the hub through which raw materials arrived and finished goods were forwarded to customers. They transformed the supply of foodstuffs to urban areas, enabling fresh produce to be transported far more cheaply and at much greater speed. They were essential in the development of modern retailing, making possible the distribution of national brands to shops in cities, towns and villages. Many large retailers depended on them into the1960s.

Goods sheds and warehouses were, in effect, the predecessor of today's big shed distribution warehouses, seen at strategic motorway intersections all over the country, and played just as significant a role in the economy as these successors do. Today they are used for a wide variety of different purposes as their large open internal spaces make them highly flexible structures that lend themselves to reuse. Although many have been demolished, they provide an opportunity for creative solutions that ensure the survivors continue to serve their communities as housing, offices, churches, museums and university buildings, just to mention a few of the uses to which they have been put.

M J Daunton

Professor Martin Daunton,
Commissioner

Preface

This book is about buildings associated with a type of railway traffic that is gone forever. Although freight haulage on railways today is booming, it is trainload traffic conveying containers and bulk haul of commodities such as coal, roadstone and oil from one terminal to another. Wagonload traffic, which is the haulage of goods by individual wagonload or less, and which had its origins in the railway's role as a common carrier (a statutory obligation to carry any type of goods), finally disappeared in 1991 but had been declining since the 1950s. In its turn, this was divided into 'full loads', which could often be handled outside in sidings, and 'smalls' or 'sundries', which had to be handled in the goods shed. Numerous small yards shut in the 1960s when traffic was concentrated at many fewer large yards. The demise of wagonload traffic was predicted in the Beeching Report of 1963, and the traditional goods sheds and warehouses seen in these pages ceased to be used as such many years ago.

A hierarchy of buildings spanned the whole country, from the multistorey warehouses in city centres, with their specialised ancillary structures such as cotton, grain, potato and bonded warehouses, through medium-sized urban goods sheds with collection and delivery services provided by the railway, transhipment sheds at junctions where goods were divided up to be transported to the far-flung parts of the railway network, and finally to the small country goods sheds or lock-up stores where shopkeepers or local merchants came with horse and cart to collect their orders. Until the 1920s, although canals retained some bulk traffic where speed of delivery was not essential and coastal shipping was also important for coal and timber, almost everything was transported by rail. Even into the late 1950s, the railways retained a substantial amount of goods traffic.

In most cases, the railways derived more of their income from goods than from passenger traffic. The proportion varied: in 1913, the North Eastern Railway, serving the mineral-rich north-east, gained 66 per cent of its revenue from goods; the London Brighton & South Coast Railway, with its heavy suburban traffic, derived just 25 per cent. Despite this, far more attention has been paid to the infrastructure catering for passengers than to that for goods and, in particular, the emphasis in the study of railway architecture has always been heavily biased towards passenger stations.

Just one general survey of goods stations has been published, a short (albeit most useful) article by Gordon Biddle in the *Railway & Canal Historical Society Journal*.[1] Very few substantial investigations of railway warehouses have been published in book form. Studies by Bill Fawcett on the facilities of the North Eastern Railway,[2] by R S Fitzgerald on the Liverpool & Manchester Railway warehouse in Manchester,[3] and by Michael Hunter and Robert Thorne on the King's Cross goods facilities[4] are some of the few works to appear, although there are a number of articles and unpublished reports on specific structures. Some 96 goods sheds and warehouses are listed or scheduled in comparison to several hundred station buildings. The subject deserves study on the same basis as textile mills, iron works, potteries and other industrial plant that contributed to the 19th-century economic dominance of Britain.

Because of the lack of study of goods sheds and warehouses, their significance has not been fully recognised and they are very much a threatened building type. Dr Michael Nevell explored the impact of development pressure on the building type and noted that of the 57 examples that had been recorded as surviving in the north-west in 1982–8, 20 (or 35 per cent) had been lost since then.[5] A similar exercise carried out for Sussex by the author between 1980 and the present revealed that of 37 examples left in 1980, 19 (or 51 per cent) have subsequently been demolished.

As major depots in cities were not open to the public and were generally hidden behind high walls, access was limited to those who worked in them or had business to transact. Outsiders were not welcome. Consequently, many goods sheds and warehouses, especially in urban areas, have been demolished completely unrecorded. Even the number of extant examples was unknown until the research undertaken for this book revealed that there were over 600 remaining in England. Other than Dr Nevell's work, there has been little attempt to create a typology for them, and a tendency to view what were essentially functional buildings in an art-historical way often leads to an emphasis on stylistic variety rather than on how they actually worked. There have been some excellent recent accounts of how the railways handled goods traffic but there has been little analysis of how this impacted on the buildings themselves.

The present book is in no sense a definitive work on the subject – that remains to be written – but is intended, within a modest canvas, to draw out the significance of railway goods sheds and warehouses as a building type, to examine briefly how they were used, to suggest a typology for their design and to look at how they may be imaginatively reused without losing all remaining internal features. The Gazetteer will, it is hoped, serve as a record of those that remain at the time of writing.

Acknowledgements

I am indebted to Simon Hickman for contributing Chapter 7 and would like to thank Kathryn Morrison, Peter Kay and Mike Nevell for reading the text. Any mistakes are, of course, my own.

The photography was undertaken by Steve Baker, Anna Bridson, Alan Bull, James O Davies, and Pat Payne. The graphics are the work of Allan Adams. I would also like to acknowledge the help of Roger Carpenter in seeking out photographs and Gordon Biddle for allowing me to reproduce several of his photographs.

Many people have contributed to the project with lists of goods sheds that they have identified, and I would particularly like to thank Gordon Biddle, Bill Fawcett, Robert Humm, Mike Nevell and the late Peter Robinson. Derek Middleton, former BR architect, gave much information about the goods sheds he worked on in the 1950s and 1960s and kindly lent a number of unique photographs for reproduction. David Challis and Mike Senatore of Industrialogical Associates gave support and encouragement, including facilitating a special feature on goods sheds in *Great Eastern Journal*. Nick Pigott, former Editor of *The Railway Magazine*, gave much encouragement and I must thank him for agreeing to print the full list of extant goods sheds in an article in the magazine. It enabled many readers to contact me following its publication, resulting in the addition of some 70 goods sheds to the record. I am very grateful to all of these readers for taking the trouble to get in touch. I would also like to thank members of the following societies specialising in the history of individual pre-1923 railway companies: the Cumbrian Railways Association; the Great Eastern Railway Society; the Great Northern Railway Society; the Great Western Railway Study Group; the Lancashire & Yorkshire Railway Society; the London & North Western Railway Society; the Midland Railway Society; the North Eastern Railway Association; the North Staffordshire Railway Study Group; and the South Western Circle.

Abbreviations

B&ER Bristol & Exeter Railway
BR British Railways
CLC Cheshire Lines Committee (GCR, GNR, MR)
CV&HR Colne Valley & Halstead Railway
ECR Eastern Counties Railway
FR Furness Railway
GCR Great Central Railway
GER Great Eastern Railway
GNER Great North of England Railway
GNR Great Northern Railway
GWR Great Western Railway
H&BR Hull & Barnsley Railway
L&YR Lancashire & Yorkshire Railway
L&BR Lynton & Barnstaple Railway
L&CR Liskeard & Caradon Railway
LB&SCR London Brighton & South Coast Railway
LCDR London Chatham & Dover Railway
LDECR Lancashire Derbyshire & East Coast Railway
LMS London Midland & Scottish Railway
LNER London & North Eastern Railway
LNWR London & North Western Railway
LSWR London & South Western Railway
LTSR London Tilbury & Southend Railway
M&CR Maryport & Carlisle Railway
M&GNR Midland & Great Northern Joint Railway (MR, GNR)
MS&LR Manchester Sheffield & Lincolnshire Railway (renamed Great Central Railway 1897)
NER North Eastern Railway
NLR North London Railway
NSR North Staffordshire Railway
PD&SWJR Plymouth Devonport & South Western Junction Railway
S&DJR Somerset & Dorset Joint Railway (MR, LSWR)
S&WR Severn & Wye Railway (MR, GWR)
SE&CR South Eastern & Chatham Railway (working union of SER and LCDR from 1899)
SER South Eastern Railway
SMJ Stratford-upon-Avon & Midland Junction Railway
WSMR West Somerset Mineral Railway

HAYCOCK & RUSSELL
COAL, CORN & CAKE
MERCHANTS
M R

1

How a goods shed functioned

The words 'goods shed' and 'warehouse' are often interchangeable in railway terminology. In their minutes or drawings, some railway companies referred to small rural structures as 'goods sheds', others to 'goods warehouses'. However, the large multistorey buildings found at major goods yards and in city centres were always only known as 'warehouses'. For the purposes of this book, the smaller structures will be referred to as sheds, the large urban multistorey structures as warehouses.

Let us begin by looking at how goods sheds and warehouses functioned. The method of operation with the smaller shed changed little from Victorian times until the 1950s, with the advent of the telephone and motor transport being the only real advances. The goods shed needs to be seen in the context of its setting. A typical goods shed would be located in a goods yard with two or three sidings. Some items, particularly bulk loads, would be unloaded outside, directly from wagons in the sidings. Part of the site would be occupied by one or more coal merchants, who would often have their own offices in small huts, and there would usually be a dock for unloading livestock. Goods traffic at local level might be handled by a checker, a porter and junior porter – the latter two would also undertake duties on the passenger side of the business and all would be under the supervision of the stationmaster. The goods shed would be open during normal working hours only – between 8 am and 6 pm. There might be one or two goods trains calling daily. Wagons for the station would be set down in the goods yard and others, either empty for return or loaded with goods, collected in their place to be sent out to the network. The shunting would be undertaken initially by the locomotive of the local goods train but, once it had gone on its way, any further movement of wagons would generally be done by a man using a pinch bar or wagon lever (a pointed steel bar inserted where the wagon wheel met the rail) or occasionally by a horse with a rope tied to a hook on the underframe of the wagon.

The operation of the goods shed was governed by paper in the form of invoices which were handed to the guard of the goods train or forwarded by passenger train. Cards (wagon labels) were attached by a clip to the solebar (side member of the underframe) of the wagon and gave details of where it was to go and the nature of the goods loaded. The destination was also often chalked on the side of the wagon. Costs of shipping goods were worked out by staff with the aid of rate books. Railway companies, as common carriers, were obliged by law

A country goods yard belonging to one of the smaller companies, the East & West Junction Railway, later the Stratford-upon-Avon & Midland Junction Railway (SMJ), at Byfield, photographed in 1904 by S W A Newton, better known for his studies of railway navvies on the Great Central London Extension. The goods shed, built in 1873 along with the station, was of a design erected elsewhere on the SMJ. The small yard was quite cramped, with coal being offloaded and stacked outside the shed, a coal merchant's office adjacent to it and a cattle dock beyond.
[BB98_06288]

Figure 1 (left)
A typical small goods shed interior. Unusually, Wellingborough MR (1857) retains its 30cwt timber jib cranes.
[DP172455]

to carry any type of traffic and quite literally thousands of rates were in force for different forms of goods. A rate would be quoted and, provided the shipper was happy with it, he would deliver the goods by cart, backed up against the internal platform of the goods shed for the items to be removed by hand and then wheeled on a trolley to await being placed in a wagon. Goods tended to be packed in wicker baskets, wooden crates, sacks, barrels or cardboard boxes. If the goods were heavy, the rotating jib crane in the warehouse would be called into action, and this could also be used when the wagons were present to lift the goods in (Figs 1 and 2). The paperwork would be dealt with in the office, which would be either within the shed or in an attached office building at one end of it. The goods staff would notify management of the amount of goods awaiting

Figure 2 (right)
By no means were all jib cranes made of timber. Here is an iron example at Congresbury, GWR, in a shed built by the B&ER in 1869.
[J H Moss/R S Carpenter Collection]

collection and how many wagons would be required to move it, and the requisite number of empty wagons would be attached to the next goods train to call at the station. Although closed vans were quite numerous, by far the largest proportion of wagons were open, and the staff would have to cover them with tarpaulins (wagon sheets) to protect goods from bad weather.

When goods arrived at the station for collection, the process was reversed. Invoices, with a description of the goods, address, weight and details of the charges, would be handed by the guard to the local goods staff. They would send a card to (or, from the 1920s, telephone) the consignee, notifying him that his goods had arrived and requesting that he collect them. In towns, there would be a collection and delivery service, operated either by the railway or by a local agent at extra cost, as an alternative to the customer delivering and collecting with his own vehicle. From the late 1920s, it was increasingly the practice to station a motor lorry at rural goods stations to make local collections and deliveries. A customer might want to keep the goods at the shed for some time, and there was often a limited amount of warehousing of this type present.

The practice at large goods stations was basically the same but on a much larger scale and with more scope for speeding up operations by mechanical handling. There were two principal differences: some goods were not just originating and terminating at the station but were being transhipped from one train to another as part of a much longer journey, and extensive space was often given over to warehousing consignees' goods (Fig 3). Specialised warehouses given over to one type of commodity – cotton, wool, potatoes, fruit, grain – or industrial products such as iron or steel sections were common. Warehouses operating under bond might be found in major cities or where dutiable products were produced or distributed. In addition, especially from the 1920s, parts of the premises were frequently sublet to outside firms for warehousing and distributing their products.

Such establishments would operate 24 hours a day, and often employed hundreds of men, including those carrying out clerical work in the offices and those loading and unloading wagons within the shed, which involved a high degree of physical labour, even with the advent of innovations such as conveyor belts. But many other occupations were represented: capstanmen, responsible for moving the wagons within the shed; men operating the hydraulic cranes and hoists; van setters, who moved the delivery vans within the depot; delivery men

Figure 3
A cutaway drawing of the Manchester, Deansgate warehouse reveals its construction with steel framing and jack arches. The first floor, with its separate access, dealt with London traffic and the ground level with other destinations, and the three floors above were used for warehousing. Among the details visible are the cages of the lifts linking the warehousing to the goods shed, the rows of cranes for lifting goods into road vehicles, the internal railed trolley system within the warehouse floors and the turntables and capstans for moving wagons.
[Allan Adams/Historic England]

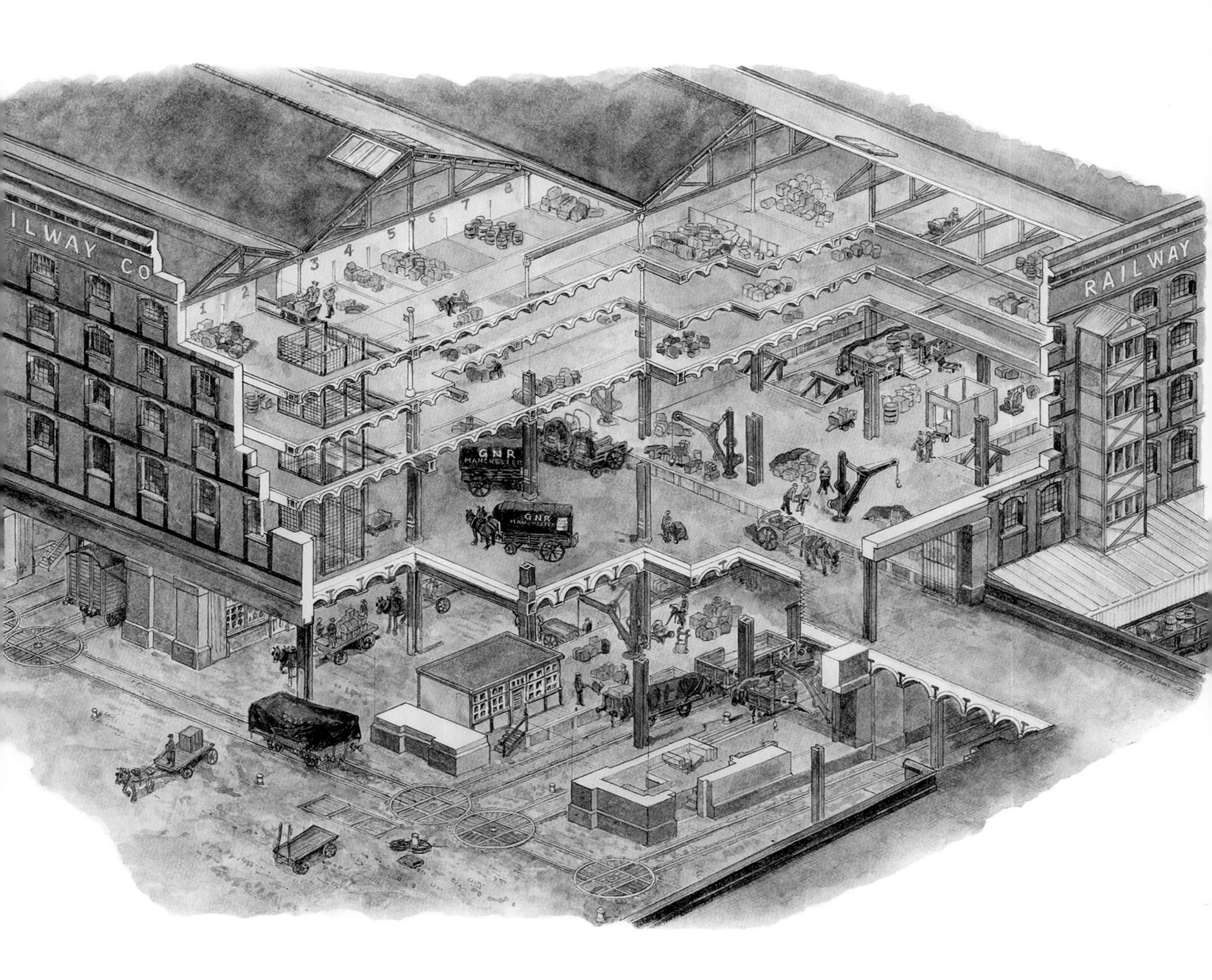
ILWAY CO
RAILWAY
1
2
3
4
5
6
7
8
GNR
MANCHESTER
GNR
MANCHESTER

Figure 4
The interior of Newcastle Forth Goods in 1894. This posed view, one of three of the depot taken by Bedford Lemere for the NER, probably to record the extension of the building by William Bell in 1891–3, gives an idea of the size of a major city goods depot and also the forms of packaging in use: the barrel, the wicker basket and the wooden crate.
[BL12500_002]

(carmen) and young boys (vanguards) assisting them, responsible for operating the extensive collection and delivery services within towns and cities; and a host of other functions including security (watchmen), stablemen to look after the horses, and teams of agents and canvassers to seek out business from local firms.

Large sheds were divided into inwards, outwards, tranship and warehouse functions. Inwards traffic, which was often dealt with overnight, was handled by invoices being entered into a book, checked and then passed to a marking clerk, so called because he marked the position on the platform where the item was to be placed. Perishables for market were given priority. The invoice was then passed to another set of clerks who entered the details onto the carmen's delivery sheets. The invoice then went to the platform, where a gang of men unloaded the wagons and moved the goods to the appropriate point on the platform. Empty road delivery vans were put in position by van setters, loaded and moved into position for the carmen to take them off early in the morning on their rounds.

For outwards traffic, the wagons that had arrived with inwards traffic were, once emptied, moved to the outwards platform. Consignment notes prepared by the sender with details of the goods, address and weight, and handed to the carman on collection, were the basic document, with a separate one for each package for a particular destination, and were stamped as they entered the goods station. The notes were then handed by the carmen to the checkers in charge of the unloading gangs, and the goods were weighed and moved by handcart to the requisite point on the platform for their destination, accompanied by the consignment note. Once the goods were in the wagons, the notes were taken to the shipping office, where the clerks made out the invoices, which were handed to the guard of the goods train containing the loaded wagons.[6] To an outsider the interiors of large goods depots could at first sight seem chaotic, with packages of all sizes and types deposited apparently at random on the platforms, but the level of control through the extensive paperwork was considerable and goods rarely went astray (Fig 4). The tranship and warehouse functions will be described in Chapter 5.

2 The origins and evolution of the goods shed

Origins

Manchester, Liverpool Road. The 1830 Liverpool & Manchester Railway warehouse as it stands today, part of the Manchester Museum of Science and Industry. [AA026851]

Figure 5 (below)
A surviving crane winch in the attic of the 1830 Manchester, Liverpool Road warehouse. [AA016862]

One aspect of railway architecture that has perhaps not received the attention it deserves is the origins of the various building types that so speedily appeared on the newly built railway system. There was clearly a precedent for the railway warehouse in the multistorey warehouses that had been constructed in the new docks of the early 19th century in London and the great northern cities such as Manchester and Liverpool, and especially those warehouses built specifically to handle canal traffic. R S Fitzgerald draws attention to the relationship between the Manchester, Liverpool Road warehouse of 1830, the first major railway warehouse to be constructed anywhere in the world, and two other structures in the city: the 1793 extension to the Grocers' Warehouse and the Merchants' Warehouse of 1825. Both incorporated transverse cross-walls which divided the buildings into a series of fireproof compartments,[7] a constructional detail shared by Manchester, Liverpool Road.

The warehouse at Liverpool Road was four storeys in height with a basement, a ground floor with road access from a yard to the north side, rail access via turntables on the first floor and a second floor used for warehousing goods. This set the pattern for subsequent railway warehouses, with few being of more than five storeys. Although turntables went out of general use for new construction after the 1860s, with access by sidings running directly into the buildings being favoured, they remained of vital importance in the cramped locations typical of many goods depots in city centres and continued to be installed in these circumstances. Internally, goods were initially lifted by gravity hoists, of which crane winches remain in the roof space (Fig 5). In 1831–5, Stephenson mechanised the lifting gear with steam power and line-shafting.[8]

A claimant to the title of the world's earliest railway goods shed is the one built by the Stockton & Darlington Railway at Darlington in 1827. Located on the side of an embankment, it was of two storeys with the upper floor at rail level and cart entrances at ground level. It was replaced in 1833 when it was converted to a passenger station and staff accommodation before finally being demolished in 1864. Some remnants of its foundations may remain on site. A similar arrangement was employed in the Leeds & Selby Railway's Micklefield goods shed of 1835 (Fig 6) and later at the Great North of England Railway sheds at Alne (Fig 7) and Thirsk in 1840. Micklefield was a plain, hipped-roof

Figure 6
A rare survivor from the Leeds & Selby Railway, the first main line railway in Yorkshire, opened in 1834, is the former goods shed at Micklefield. Its present appearance dates from 1886, when it was converted to a house and a large central window was added.
[DP169054]

Figure 7
Alne goods shed, now disguised as a terrace of houses. It was built for the Great North of England Railway in 1840 and was one of the first goods sheds to be designed by G T Andrews, with his favoured details of round-headed arches and a hipped roof.
[DP169010]

brick structure with small-paned iron windows, and rail access probably centrally via a turnplate to the upper floor. It survives, with alterations resulting from its conversion to a house in 1886, to give an idea of the layout of the now lost first goods shed at Darlington.

The 1827 Darlington shed's 1833 replacement, the Merchandise Station (Fig 8), still stands, the earliest single-storey goods shed to survive. Designed by Thomas Storey, Engineer to the Stockton & Darlington Railway, the Merchandise Station was extended in 1839–40 when a handsome Classical clock tower was added. The layout did not influence the future design of goods sheds in that access for rail traffic was via a series of short sidings entering the building at intervals along its principal façade rather than a line running through the shed. Constructed of sandstone, it comprises eight bays separated by pilasters and with large round-headed iron windows.[9] Its Grade II* listing reflects its substantially unaltered condition.

Figure 8
The second-oldest surviving goods shed in England, the Darlington Merchandise Station of 1833. The clock tower, added in 1840, with its finely modelled Classical detailing, gives it something of the air of a country house stable block
[DP169016]

Canal warehouse inspiration can be seen in a number of warehouses built in the early 1840s, of which Heywood (1841–3) (Fig 9) is a notable survivor. Built for the Heywood branch of the Manchester & Leeds Railway, it had a track running through it and a warehouse section of two storeys at one end, with similarities to a now demolished canal warehouse nearby. Contemporary drawings indicate that up until the mid-1840s, and in some cases beyond then until *c* 1850, goods sheds were accessed via turntables rather than by direct track access. The earliest surviving sheds at Darlington (see Fig 8), Micklefield (see Fig 6) and Hexham (Fig 10) are all of this type. S C Brees, in *Railway Practice, Fourth Series* (1847), illustrates a number of small station layouts including those at Tring, Wolverhampton and Woking, all of which have this arrangement. Many were built at right angles to the tracks rather than parallel to them. There are a number of surviving examples of this, among them Paddock Wood, Wareham (*see* Fig 72), Hassocks, Wymondham (*see* Fig 45) and Ashwell (Rutland). The practice was not universal, as Brunel – individualistic as always – favoured goods sheds built parallel to the tracks. On the Bristol & Gloucester Railway, this approach may still be seen at the shed at Yate (1844), which, like the other wayside stations

Figure 9
Heywood (1841–3). An early shed constructed by the Manchester & Leeds Railway at what was then the terminus of a short branch, subsequently extended to Bury.
[DP169308]

Figure 10 (above)
The interior of the Newcastle & Carlisle Railway goods shed at Hexham of 1835, with its ridge-and-furrow roof.
[DP169022]

Figure 11 (above right)
The 1838 goods shed at Fourstones, Newcastle & Carlisle Railway.
[Bill Fawcett]

on the line, was entered via a turntable into the end elevation (*see* Fig 32). The Newcastle & Carlisle Railway shed at Fourstones (1838) (Fig 11) may also have been of the parallel type.

Location in cities

The location of goods facilities was an important factor in determining their layout and subsequent growth. At Manchester, Liverpool Road, the goods warehouse was located immediately opposite the passenger station but, before long, the separation of large goods stations and passenger stations became the norm. At Liverpool, the Crown Street passenger station was quite distinct from the goods yard, which was located to its north. The London & Birmingham Railway did not transport goods south of Camden where extensive goods yards were laid out (*see* Fig 51), passengers travelling the last mile down to the Euston passenger terminus by gravity. Similarly, at King's Cross on the Great Northern Railway, the goods yards were to the north of the Regent's Canal. Some depots, such as those at Manchester, London Road (LNWR/MSLR, now Piccadilly) (*see* Fig 58), Manchester Central (CLC/MR), Paddington (GWR) (*see* Fig 64) and St Pancras (MR), were adjacent to the passenger termini but there was a noticeable separation between them. All the railways serving the area south of London had goods facilities far removed from their passenger stations – hence the LB&SCR goods equivalent of Victoria was at Battersea, the LB&SCR goods equivalent of London Bridge was at Willow Walk, the SER goods equivalent of Charing Cross was at Bricklayers' Arms, the SER goods equivalent of Cannon Street was at Ewer Street and the LSWR goods equivalent of Waterloo was at Nine Elms.

In many cities, whole areas became dominated by extensive complexes of railway yards and warehouses, sometimes raised up above them. In many cases, these complexes were formed around former passenger stations that became too small to deal with traffic growth, were by-passed by later extensions of the system or were simply found to be located in the wrong part of the city. Manchester, Liverpool Road owes its survival in so intact a state to this. Examples are legion: Manchester, Oldham Road (formerly the Manchester & Leeds Railway terminus); Bradford, Adolphus Street (Fig 12); Nine Elms; Birmingham, Curzon Street (*see* Fig 57); London Bishopsgate (Fig 13); Leeds, Marsh Lane; Sheffield, Bridgehouses and Wicker, to name but a few.

The depots could form clusters in certain locations, as each competing railway company wished to ensure that it had a share of the trade. One of the most striking of these is the concentration of depots on the final stretch of the GER line running into Fenchurch Street station. Within less than a mile, there were the following depots: LNWR Haydon Square (1853); GER Goodmans Yard (1861); MR City Goods (1862); Royal Mint Street (built 1858, leased to GNR from 1861); GER East Smithfield (1864); and LTSR Commercial Road (1886). The same was to be found in many dock areas: at Poplar Docks, the GWR, GNR, LNWR and NLR had depots, each accompanied by a large warehouse. The textile towns of the West Riding saw something similar. At Halifax, there were separate warehouses that the L&YR shared with the GNR, the LNWR and the MR respectively (*see* Fig 50). At Dewsbury, the LNWR, the GNR, the L&YR and the MR all had separate establishments, almost encircling the town centre.

Another factor in determining warehouse location was proximity to waterborne traffic. Besides the presence of railways in docks, there were numerous instances of goods yards being constructed with canal or inland

Figure 12
Bradford, Adolphus Street on 14 July 1937. At the top right is the former passenger station, opened in 1854 and converted to a goods depot in 1867. Below it is the original goods shed, with covered loading platforms on each side and used for fruit traffic. On the left are two massive three-storey wool warehouses, one a triple-pile structure of 1873 and the other a later addition with a ridge-and-furrow roof.
[EPW054318]

navigation basins to enable transhipment to take place from water to rail. For example, the basin provided at the new GCR goods warehouse at Lincoln (Fig 14) in 1907 allowed boats from the River Witham to discharge cargoes directly into the warehouse or railway wagons. Another late example of this was at Manchester, Deansgate (*see* Figs 3 and 63), where a canal basin was excavated below ground level. Other examples of such facilities were a covered wharf adjacent to the Regent's Canal at Marylebone GCR (*see* Fig 62); a link to

Figure 13
Bishopsgate, the GER's principal London goods depot, built on the site of the original Eastern Counties Railway passenger terminus. The sheer scale of a major London goods depot is apparent in this view of 30 September 1947, looking north. It was built in a single phase in 1878–81 with a large part of the site under cover, the architect being the GER's Edwin Wendover.
[EAW011220]

Figure 14
The former Great Central Railway warehouse (1907) at Lincoln, now converted to the library of the University of Lincoln.
[DP173937]

the Nottingham Canal at Nottingham MR; a wharf at Bristol, Temple Meads GWR; and direct underground access from the Regent's Canal to the LNWR Camden goods warehouse (*see* Fig 51) and to the GNR King's Cross granary (*see* Chapter 7 opener). Even earlier was the London & Birmingham Railway goods warehouse at Wolverton, situated next to the Grand Junction Canal, which had cranes on its internal platforms to lower goods to the canal, with the roof projecting 'over part of the canal to protect the barges in bad weather'.[10] Thus, with rail, road, and waterborne traffic all in play, the major goods station was a multimodal transport hub as far back as the mid-19th century.

Architectural style

Goods sheds were seen from the start as functional structures having more in common with other types of warehouse than with adjacent passenger stations. Nonetheless, there was a definite attempt in some cases to produce designs that were simple but elegant. The Darlington Merchandise Station of 1833, with its passing resemblance to a stable block, has already been noted. Francis Thompson in his goods sheds at Ambergate (Fig 15) and Matlock (1849) used a well-proportioned hipped-roof design in fine ashlar. The large goods sheds of G T Andrews, for the Great North of England Railway and other companies controlled by George Hudson, formed part of a clearly expressed architectural vocabulary in which the various buildings, goods and passenger alike, complemented each other (Fig 16).[11] The inspiration for Andrews's designs was model farms and stables: their proportions, the use of a stone plat band around the buildings, and hipped kingpost roofs display this influence. Andrews's trademark Diocletian windows may well have been derived from farm buildings and stables designed by Sir John Soane, many of which incorporated this detail, for example those at Cadlands Farm (1777) and at Betchworth Castle (1798–9). These buildings by Andrews reflected the trend towards what J M Richards named 'the functional tradition'. It led to relatively plain designs with the odd flourish, such as the MR grain warehouse at Helpston, with its shaped timber eaves brackets (Fig 17); the B&ER goods sheds at Congresbury or Axbridge, with elegant cast-iron Gothic or Florentine windows (*see* Figs 2 and 37); and the series of goods warehouses, lit by lunettes, on the LB&SCR, described by Ian Nairn as being 'still Georgian in proportions and delicacy' (Fig 18a and *see* Fig 28).[12]

Figure 15
Ambergate, a simple but elegant design of 1840 by Francis Thompson, for the North Midland Railway, photographed in March 2001.
[Gordon Biddle]

Figure 16
The characteristic type of goods shed designed by G T Andrews for companies that became constituents of the NER, this one at Hutton Cranswick. The very broad Diocletian windows linked by a band at sill level, round-headed openings and the low hipped roof all show a possible influence from Sir John Soane's designs for stables. Exceptionally large and handsome structures, they have lent themselves to re-use, this one being used as offices.
[DP169056]

Figure 17
The grain warehouse at Helpston, MR, seen in 1957. The brackets turn a utilitarian structure into a building of considerable visual appeal.
[Gordon Biddle]

Figure 18
A series of detail views that help define the goods-shed aesthetic:
(a) Arundel (1863) [DP172806];
(b) Langwathby (1876) [DP169032];
(c) Stafford (1880) [DP173404];
(d) Singleton (1881) [DP172827];
(e) Broadway (1904) [DP173379].

Deliberate attempts to match buildings to neighbouring passenger stations were less common. In his use of Tudor Gothic, Brunel attempted to match goods sheds to station buildings in the same style as at Culham or Yate (*see* Fig 32), but he frequently resorted to simple timber structures. The distinctive FR goods sheds of the 1860s to 1880s employed the same vivid polychromy as the Paley & Austin-designed stations of the period. MR goods sheds on the Settle & Carlisle and the Nottingham–Worksop lines, opened in the 1870s, had similar windows to their passenger counterparts, with pointed arches and cast-iron window frames (Figs 18b and 19). The most extreme instance of buildings designed by an architect to match station buildings involved three goods sheds designed for the LB&SCR by T H Myres in domestic revival style with roof tiles rather than slates, terracotta ridge tiles and half-timbered panels with incised flower decoration – Singleton (1881, listed II) survives (Fig 18d). The level of external decoration varied considerably from company to company, but commonly included polychromy or brick nogging at the top of recessed panels, along the eaves or in the gables. Some companies, such as the B&ER and the MR, favoured bargeboards. Several had circular openings in the end gables to provide through ventilation.

Figure 19
Langwathby, one of the standardised designs erected by the MR on the Settle & Carlisle line in 1876, showing the high-quality masonry, the use of pointed arches and elaborate cast-iron window frames, particularly associated with this line but also used elsewhere by the MR. The angled boarding on the doors was another original detail of these buildings.
[DP169031]

Materials and design details

Goods sheds were generally built of local materials, especially up to the 1870s. In stone districts, ashlar was generally left rock-faced, as in the early L&YR shed at Cherry Tree (Fig 20). By the 1870s, bricks were often no longer locally made but imported from elsewhere: the GWR used hard red Ruabon brick and the LNWR established its own brickworks at Crewe in 1862 to supply the whole system. Common to all methods of construction was the practice from the 1860s, and in many cases earlier, of division of the shed into clearly defined bays. In brick, this was achieved by constructing a series of recessed panels, often flanked by what were, in effect, pilasters. The panels could be flat-headed or with segmental or pointed arches. This form of construction was for economy: it saved considerable numbers of bricks. In stonework, the same visual effect was obtained with buttresses and, in timber, by division by vertical posts.

Figure 20
An early (1847) and characteristically massively built East Lancashire Railway goods shed at Cherry Tree, Lancashire, a wayside station on the Preston–Blackburn line.
[DP169045]

Timber was also widely used. In some cases, in the early days of railways, it was uncertain to what extent the traffic would develop and a timber structure was put up which could later be replaced with a larger masonry building if required. But timber was always a popular material and was often used where ground conditions would not support a heavier structure, where subsidence (especially in coal mining districts) was a problem, or for the internal platform within a masonry building. Some companies made very extensive use of wood: until the 1860s the ECR employed it for many of its sheds, including some of considerable size, while the Highland Railway used it for almost all construction for the whole of the company's separate existence. There was something of a revival of timber construction in the late 19th century when many companies sought to make economies and small, rather rudimentary timber structures were erected by several companies, notably the MR and the LNWR. Both companies relied on skylights for lighting these buildings. Framing generally consisted of a box frame with the building divided into bays by vertical posts at regular intervals, with horizontal rails at the base and top of the wall, and a third rail running along the building halfway up its height. The frame thus formed was then braced by X-shaped members in each bay. The cladding of timber goods sheds could vary greatly: the boarding could be vertical or horizontal, weatherboarded or board-and-batten, and while most were creosoted, some were painted. Corrugated iron began to be used for some small structures by the late 19th century.

Goods sheds were fitted with double doors across the track at both ends. Generally, these were of full height but occasionally half-height ones, rather like those found in a Western saloon, were used. Such doors were always liable to damage and, in later years, many small rural sheds lost them. Doors were usually hinged, but sliding doors, with small wheels located in iron runners, were frequently used on the cart entrances, placed either inside or outside the building and, less often, across the track.

Windows varied considerably in their number and position. Many companies were content to light their goods sheds entirely by roof lights running much of the length of the shed just below the gable. This had the great merit of ensuring security. Some other companies favoured small windows relatively high up in the side walls of the structure, lunettes being particularly favoured, as security was still not compromised. In large goods sheds in urban areas, windows in the sides of the building supplemented roof lights and tended to be larger in size and closer to the ground. However, such buildings were located in secure areas, enclosed by high walls with a single exit guarded night and day, reducing the possibility of the building being broken into. Even in such areas, however, warehouses still had relatively small windows, probably because there was not the same need for light as there was in a goods shed where goods were continually being unloaded and transhipped, where consignment notes had to be read and where there was constant movement with barrows and trolleys. Bonded warehouses had very small window openings, in accordance with regulations relating to bonded storage.

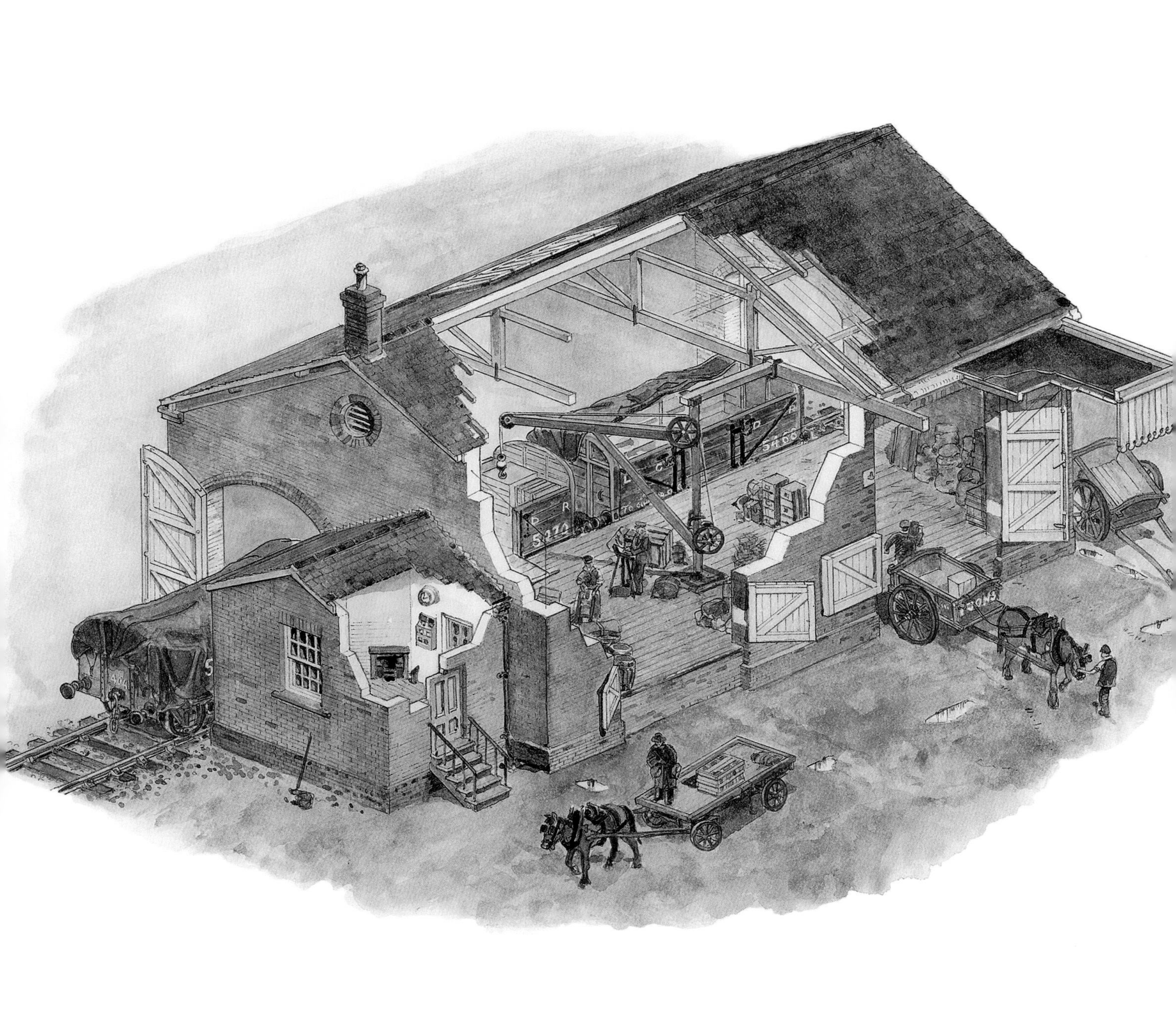

3 Plan forms

Goods sheds varied greatly in size, from small lock-up structures serving villages to multistorey warehouses in city centres. The following is an attempt to establish a typology of the various designs (Fig 21).

1 The through type

These were by far the most common type of goods shed and were to be found in both urban and rural locations, varying greatly in length. They had a single track running through them and all but the smallest had an internal platform of timber, brick or stone which was generally equipped with a timber jib crane, typically with a capacity of 30cwt, that rotated from a top bearing placed in a roof truss and a lower bearing located in the platform. The basic design was already in place in the 1840s. A great many variations were possible within the basic design and these tended to revolve around the platform arrangement, access for cart loading and the provision and position of the goods office. Many of the general comments on design apply equally to the other types described here.

Access for customers' carts was via doors in the side elevation of the building. In many cases, the platform had cut-outs in it into which carts could back for loading under cover (Figs 22 and 23). One approach to this was to have canopies over the doors; alternatively, the roof could extend down to serve the same function. Canopies could be individual to each opening or continuous, protecting the whole side of the shed from the elements. An otherwise typical example of the type at Calveley had an unusually deep canopy extending over the Shropshire Union Canal, in recognition of its role as a transhipment shed between rail and waterborne traffic. In some cases part of the internal space might be partitioned off to provide an office, usually with glazed timber screens to give greater light in the somewhat dark environs of the shed. An alternative approach was to have a separate goods office as an adjunct to the building, usually attached at one end of it. These were generally small, lit by a single fire, and sometimes in contrasting material to the main part of the shed. They were in some cases later additions. Larger examples of sheds could have substantial brick-built offices, sometimes of two storeys, housing many clerks (*see* Fig 52).

A cutaway reconstruction, based on original contract drawings, of the goods shed at Rainham, LC&DR, built in the early 1860s. The shed is typical of by far the most common type of small goods shed and displays many characteristic features: the timber platform, lighting by rooflights, circular end ventilators, the 30cwt jib crane built into the roof trusses, the whitewashed interior, the heated goods office with doors leading both into the shed and externally and the canopy to shelter the loading of carts. Similar sheds were put up at various locations on the LC&DR: Rainham has been demolished but the almost identical shed at Teynham survives. [Allan Adams/Historic England]

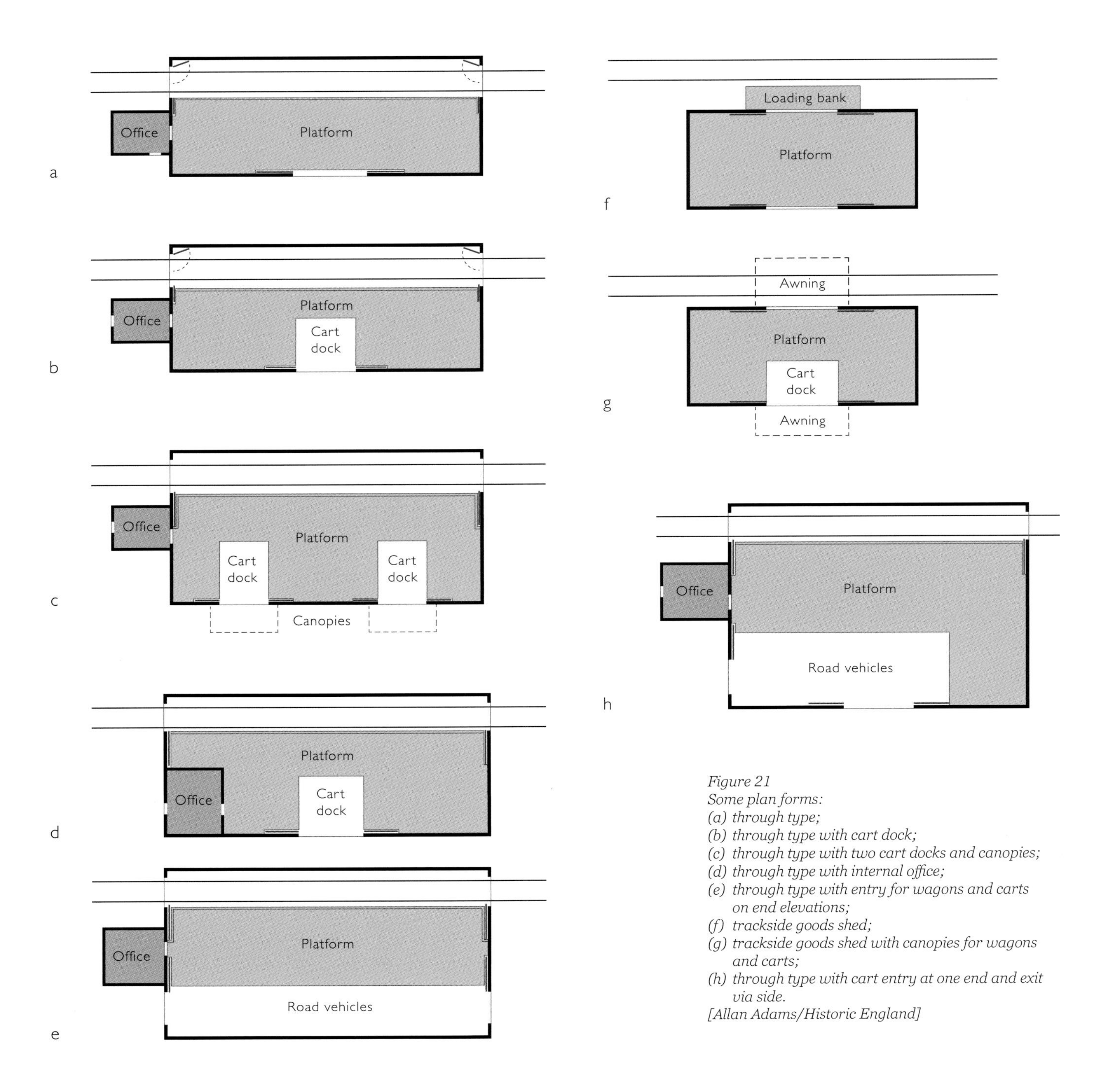

Figure 21
Some plan forms:
(a) through type;
(b) through type with cart dock;
(c) through type with two cart docks and canopies;
(d) through type with internal office;
(e) through type with entry for wagons and carts on end elevations;
(f) trackside goods shed;
(g) trackside goods shed with canopies for wagons and carts;
(h) through type with cart entry at one end and exit via side.
[Allan Adams/Historic England]

Figure 22 (above)
An indent in a platform for carts to be loaded at Ottery St Mary, LSWR (demolished). Note the timber jib crane fixed at the junction of the roof trusses, and the externally hung sliding doors for the cart entrance.
[John Eyers Collection/ South Western Circle]

Figure 23 (above right)
An example of a through shed: Ashburton (1872), designed by P J Margary, Engineer to the independent Buckfastleigh, Totnes & South Devon Railway. It had an internal office, its position marked by the chimney, and a particularly broad cart entrance with doors sliding on external iron runners.
[DP167236]

2 Through type with entry for wagons and carts on end elevations

This type, with a through cart road, was relatively early, with its origins in the goods sheds built by Brunel in the 1840s. The advantage was that the horse and cart could simply be driven through the shed and stand alongside the platform to be loaded, with no awkward reversing of the cart into confined cut-outs in the platform. The disadvantage was that it all took up much more space that could otherwise be occupied by a larger platform with more space to sort and store goods.

3 Entry at one end and exit via side

In a variation of type 2, there was an entry for both carts and wagons at one end of the building but, instead of exiting at the opposite end, carts left via an opening in the side – rather like a combination of types 1 and 2. This type was quite rare and was early: examples have been noted at Summerseat (1846) and Twyford (1857) (demolished). The drawback was that it gave little turning room and reduced platform space.

4 Wagon entry via turntable through side

These sheds were entered via wagon turntables through an opening in the side of the building. This is another early arrangement often used where goods sheds were not parallel to the tracks but at a 45- or 90-degree angle. Some sheds of this type were later rebuilt to a more conventional arrangement with through tracks.

5 Open or semi-open shed

A further variation found in the pre-1850 period was open or semi-open sheds, that is, where one or more walls were entirely open or only the top part was clad. This tended to be employed in sheds built entirely of timber, but Wootton Bassett GWR was of brick and timber construction. Brunel's brick-built sheds on the Bristol & Gloucester Railway were also largely open on the cart-loading side. Many of G T Andrews's sheds – for example, Thorp Arch – incorporated an open section at one end with the roof supported on iron columns. Although open-sided sheds tend to be confined to the early period, large examples were greatly favoured into the 20th century for loading and unloading fruit, vegetables and potatoes in major goods depots.

6 Trackside goods shed

These were small sheds, placed alongside a siding with no track running within the building. Instead, there was often a loading platform in timber or brick alongside the siding between it and the building or, alternatively, the wagon may have been unloaded directly into the shed. Some such sheds had a canopy extending over the wagon to keep those carrying out unloading operations dry and, in some instances, an equivalent canopy on the other side for the cart loading. A design of this type, which had the roof extended on both sides to form canopies and an interior lit only by a roof light, was favoured by the LNWR in the late 1890s and early 1900s.

7 Lock-up shed

These first appeared in the 1880s, serving as purpose-built structures in locations where traffic did not justify a full-size goods shed or, alternatively, erected on station platforms to supplement existing facilities and handle goods delivered by goods train to the passenger platform. Prior to that time, it appears to have been the practice to employ a life-expired passenger carriage or goods van body located either in the goods yard or on the station platform. Not all companies used lock-up goods sheds. Among those particularly favouring the type were the LB&SCR, the SER, the NER and the GWR. The LB&SCR put up more than 30 lock-up sheds, the earliest being at Pevensey in 1888. They were timber, with monopitch roofs, built usually by the company's own staff and with a sliding door on each of the principal elevations and small windows in the ends (*see* Fig 29). No two were exactly alike. The SER favoured what it termed 'roader' sheds, located on station platforms and with a pitched roof with a gable facing the platform edge. The NER favoured a similar design, although a little larger; a well-maintained example survives at Levisham. The GWR too favoured platform-mounted sheds in the early 1900s, built substantially in brick, with hipped, slated roofs.

8 Two-storey combined goods shed and warehouse

Larger goods sheds might be of two storeys with an upper floor providing space for warehousing goods. Access to the upper floor was by traps in the floor through which goods could be winched, taking-in doors with a winch above them or, in the largest structures, lucarnes. Such two-storey goods sheds could vary considerably in size, from relatively small buildings such as that at Arundel (*see* Fig 18a) to very large examples such as Bedford (*see* Fig 52), where a double-pile arrangement was employed with extensive use of structural steel and iron columns. In some instances, such as in the spinning and weaving districts of Lancashire and Yorkshire, the buildings could even be of three storeys, such as the Waterfoot goods shed (Fig 24).

9 Multistorey warehouse

The multistorey goods warehouse had a transit role in that the ground floor was devoted to incoming and outgoing goods but a key function was also to store goods. Some warehouses had internal hydraulically operated wagon lifts to enable wagons to travel throughout the upper floors via a network of turntables for loading and offloading. Such buildings, discussed in Chapter 5, were generally located within major cities.

Figure 24
Waterfoot, on the now closed line between Rawtenstall and Bacup, typical of the massive goods sheds put up by the L&YR to serve extensive local industry. Note the openings for winches to the upper two floors and the height of the lower floor windows from the ground to deter thieves.
[DP169052]

10 Stub-type shed

From the beginning of the 20th century, there was a change in the design of major goods sheds to a system of short stub sidings running into bays, each cutting into a platform area, rather than long through roads. An example was the now demolished Willow Walk of 1903 (Fig 25) where separate platforms served a limited range of destinations.

11 Steel-framed open shed

Following World War I, there was a move to much simpler goods stations which comprised very large open sheds, steel-framed with corrugated iron cladding, and with the traditional fixed cranes, which restricted movement within, replaced by overhead and mobile cranes. Warehousing was now often carried out in separate concrete or steel-framed multistorey buildings.

Figure 25
Willow Walk, LB&SCR. The major extension of 1903, showing the arrangement of short bays projecting into the loading area, with each bay handling traffic for a named destination. The photograph was taken that year as part of a fact-finding mission for the Caledonian Railway, which was investigating how various English railways handled goods traffic.
[John Minnis Collection]

4 Company designs

Goods sheds can be categorised in a number of ways – there are designs associated with a particular line or with a particular engineer, and there are those that are standard designs of a particular railway company. It is often assumed that the latter date from the 1860s onwards but, although standardisation certainly increased at that point, it had its origins much earlier. However, the absence of detailed studies of the building type on a company-by-company basis makes it hard to generalise, and this chapter just gives a brief overview of developments for some of the major railways.

London Brighton & South Coast Railway

One railway may serve to give a more detailed picture of how design evolved. The company concerned, the LB&SCR, derived the smallest proportion of its traffic from goods traffic of any of the major British railway companies but its buildings nonetheless exhibit all the stages of development, with the exception of the largest multistorey warehouses, and its designs had a higher than average architectural standard.

The LB&SCR was, as mentioned, primarily a passenger line, deriving the bulk of its income from the highly lucrative traffic from London to the south coast seaside resorts and Portsmouth and from its extensive suburban network. Its goods traffic consisted of wagons arriving on its system by the West London line via Kensington Olympia, through the Metropolitan Widened lines via Farringdon and through the Thames Tunnel via New Cross. Its London goods depots were at Battersea Wharf (West End) and Willow Walk (City), just off the Old Kent Road, and it had its own dock at Deptford Wharf. Besides general products, incoming traffic was mainly house coal. Traffic generated within the LB&SCR's territory was largely agricultural produce, timber and livestock, with some manufactured goods from the Brighton area and from south London. In addition, there was continental traffic, much of the inwards traffic being perishable foodstuffs, imported through Newhaven.

Most of the company's early goods sheds were small structures, built in either brick or timber. They had no windows, lighting being by skylight, and were generally of three or four bays. Surviving examples include Ewell (1851) and Angmering (1853). Sliding central doors for cart access were fitted beneath

The goods shed at Winchfield, Hampshire (now demolished) was built in 1904 when the LSWR main line from Woking to Basingstoke was widened, resulting in the partial rebuilding of the station. The design, with its broad three-light window in the end gable and canopies over the cart loading bays formed by continuing the slope of the roof, is characteristic of LSWR practice in the late 19th and early 20th century. [John Eyers Collection/South Western Circle]

cast-iron lintels. Similar buildings, albeit with hipped rather than gabled roofs, continued to be built into the 1860s – Hartfield (1866) and Baynards (1865) are the sole remaining examples.

Cast-iron lunettes, mounted high, first made their appearance in 1853 at Hassocks and at Earlswood, and were to be a feature of LB&SCR goods sheds for the next 30 years. A variant with a hipped rather than a gable roof opened in 1859 at Billingshurst and Pulborough (Fig 26) on the Mid Sussex line. The high point in this style was reached with a group of two-storey structures incorporating warehousing space on the upper floor. Five were built at Littlehampton, East Grinstead, Arundel, Bognor and Seaford, with Arundel (1863) (*see* Fig 18a) being the sole extant example. The central bay on the road side is gabled and incorporates a taking-in door and a winch above. The goods office is external, unlike many earlier examples, giving more space within.

The design was further refined in the 1880s with some large warehouses erected at country stations, the three survivors being at Chichester (1881),

Figure 26
A goods shed, integrated with a passenger station and with an awning on the platform side, at Pulborough. The shed, which, together with the adjacent station buildings, dates from 1859, incorporates the lunettes favoured by the LB&SCR over many years. The building's present use relates directly to its location, housing a business servicing automobiles while their owners are at work, ready for them to collect when their train returns in the evening.
[DP172826]

Edenbridge (1888) (Fig 27) and Burgess Hill (1889). These had some restrained polychromy, brick cogging below the eaves and, for the first time, included canopies on the road side to provide shelter to the waiting drays. They were built in some quantity and became the company's standard design, employed into the late 1890s. The 1880s also saw what was perhaps the most ornate goods shed ever built in this country. A number of country branch lines were opened with large station buildings designed in Domestic Revival style by the architect T H Myres. Myres also designed five goods sheds to accompany them, of which only one survives, at Singleton (1881) (*see* Fig 18d). The goods sheds picked up many of the decorative themes present in the station buildings, the most striking being applied half-timbering and plaster incised with fashionable icicle motifs. In addition, the roofs were tiled rather than slated, and had terracotta ridge tiles. Windows in the external goods offices were of the 'Board School' type with small paned upper sashes over plate glass lower ones. The basic shape of the goods sheds resembled that of earlier LB&SCR ones,

Figure 27
Edenbridge Town (1888) is an example of the LB&SCR's impressive standard design, which was to be found throughout its system.
[DP166733]

but the decoration added in keeping with the nearby stations made them unrecognisable as such.

In complete contrast, by the end of the 19th century economy had become a major consideration. Goods sheds lost the panelled wall surfaces, the lunettes and decorative brickwork and instead had rectangular steel windows, decoration being confined to a lunette in the gable end. Four of this type survive: Hayling Island (1900), Christ's Hospital (1902), Bognor (1903) (Fig 28) and Crowborough (1906). In addition, some even more austere steel-framed goods sheds with barrel-vaulted roofs and corrugated iron cladding went up from 1898 with Emsworth (1912), the last to go in *c* 1985.

Figure 28
Bognor Regis (1902) had a very large goods shed erected as part of a comprehensive rebuilding of the station. The elaboration seen in earlier LB&SCR sheds has disappeared, although the valence to the canopy injects a little decoration. As befits its size, with no less than five cart entrances, necessary for a rapidly expanding seaside resort, it has a substantial two-storey goods office. Its use as a builders' merchants has led to it remaining in largely original condition. [DP164894]

From the 1880s, there was a need for covered accommodation at many of the LB&SCR's smaller rural stations, and the company provided a structure known as a goods lock-up, a small timber building which was usually erected at a cost of about £20 by the company's men as opposed to a building contractor. No two were exactly alike but they had a family resemblance, with a monopitch roof and horizontal weatherboarding. Sometimes windows were fitted, sometimes a small loading platform was installed and a few had a simple valance around the eaves. Over 30 were constructed between 1888 and 1900: Horsted Keynes (1890), now moved to Ardingly, and Isfield (1898) (Fig 29) are well-preserved survivors.

Figure 29
Isfield, one of the smallest types of goods shed, called a 'lock-up goods' by the LB&SCR, was built in 1898. The ramp and canopy were added after the site became a heritage railway.
[DP172818]

London & South Western Railway

Line styles were much in evidence on some parts of the LSWR. The Salisbury to Exeter main line had many examples of the same type of shed, such as the neo-Gothic design once found all along the line from Yeovil to Exeter, probably the work of Sir William Tite's office, with surviving examples at Broadclyst and Crewkerne (Fig 30) (both of 1860) or the simpler brick structures on the Salisbury–Yeovil stretch, of which the sheds at Semley and Templecombe (also of 1860) survive.

Two designs by W R Galbraith, the LSWR's retained Engineer for New Works, for a simple trackside stone shed – one small and the other slightly larger with a heated office partitioned in one corner of it – were to be found throughout the North Cornwall line, with extant examples at Port Isaac Road, St Kew Highway (1895) and Tresmeer (1892), but they were confined to that line rather than becoming a system standard, with just one further example at Corfe Castle (1885).

Standardisation to some extent was to be found in a series of large brick goods sheds put up in the 1880s and 1890s with a broad three-light window in

Figure 30
Crewkerne (1860), one of a handful of remaining examples of the standard design prepared by Sir William Tite's office for the LSWR Yeovil–Exeter main line. Broadly Gothic in common with the station buildings on the line, it has corner buttresses and blind arcading on the elevation facing the railway. [DP172852]

Figure 31
Wool. A trackside goods shed of a standardised LSWR design.
[DP166741]

the gable ends as the key distinguishing feature. Examples included Brockenhurst, Farnham and the now-demolished sheds at Brookwood, Winchfield and Walton-on-Thames. Another type in this category was a small lock-up design with a hipped roof, still to be found at locations as far apart as Wool (Fig 31), Witley and Hampton Court.

South Eastern & Chatham Railway

The two companies, the SER and the LCDR, that joined together in a working union in 1899 brought a wide variety of different goods shed designs to the enterprise. Early SER goods sheds were plain brick structures, sometimes at right angles to the running lines, with exposed rafter ends on the facia boards of the gables. One, possibly of 1842, survives at Paddock Wood. The emphasis was more on designs associated with particular lines: a distinctive design with slit-like windows was to be found on the Tunbridge Wells–Hastings and Hastings–Ashford lines, opened in 1851–2; Robertsbridge and Appledore survive. By contrast, the Maidstone–Paddock Wood line of 1844 had ornate sheds of the through-cart-road type with lunettes and kneelers, very similar to those being built by the GNR at the time, with a larger example of 1846 at Canterbury West. Some larger warehouses were built for hop traffic at Tonbridge and Staplehurst. The LCDR sheds, similarly, were a mix of the relatively plain, such as at Adisham (1861) and Teynham, and rather more ambitious structures such as that at Herne Bay (1861), which originally had a half-hipped roof.

Great Western Railway

As ever, I K Brunel was an innovator in this field. Just as he had standardised station buildings, he produced a number of standard goods shed designs. These were built in timber, stone and brick or a combination thereof. There is a clear distinction between those that were relatively plain structures, most of which were of timber construction or had timber end walls, and the much more elaborate examples in Tudor Gothic style which were intended to harmonise with Brunel's stations in the same style.

Taking the latter first, they were some of the most architecturally ambitious of early goods sheds. Roof pitches were steep, gables were surmounted with shaped mouldings, a small decorative pointed arch was set into the gables and the corners were provided with buttresses of distinctly ecclesiastical appearance. A track ran through the shed and access for carts was via a single opening with a four-centred arch. Known examples include Culham (1844), Wantage Road, Southall, Henley (believed to have been relocated from Twyford) and Yate (Fig 32), the latter built for the Bristol & Gloucester Railway (1844) which was absorbed into the MR. These were all small structures, no more than two or three wagons in length. Most survived into the 1960s but Yate is the sole remaining example. Much larger versions were erected at Challow and Stroud.

The timber goods sheds were numerous and although, like the Tudor Gothic designs, many lasted into the 1960s, they are now extinct.[13] They were built along the Berkshire & Hampshire Railway at Theale, Aldermaston and Thatcham (all of 1847), with a series on the Oxford–Worcester line, opened in 1852–3, at Hanborough, Charlbury, Shipton, Adlestrop (Fig 33), Moreton-in-Marsh, Chipping Campden, Evesham and Pershore. All were clad in shiplapped

Figure 32
Yate (1844), the only survivor of Brunel's small Gothic goods sheds put up at wayside stations, this one located ironically on an ex-MR main line rather than one belonging to the GWR, because of Brunel's role as Engineer to the Bristol & Gloucester Railway, an MR constituent.
[DP172839]

Figure 33
Adlestrop (1852). A Brunel-pattern timber shed of the through type, with cart entries on the end elevations, built for the Oxford, Worcester & Wolverhampton Railway, in its last days in the 1960s. The lean-to goods office is original, its monopitch roof typical of sheds built for the GWR and constituent companies prior to the 1870s. The large rounded opening for carts is clearly visible.
[R G Nelson/R S Carpenter Collection]

timbering and were of the through roadway type with twin rounded openings in the end elevations.

A development of the design had brick or stone side walls with the ends in shiplapped timber, again generally with through roadways and openings at each end. Some large and imposing examples of this type were built at Newbury and Chippenham in 1856–7 and at Maiden Newton and Dorchester on the Wiltshire, Somerset and Weymouth line, opened at the same time. One example of the type at Ross-on-Wye (1855) is still extant (Fig 34).

Another variant of Brunel's designs can be seen in his sheds for the transhipment of goods from the narrow to the broad gauge and vice versa, which reveal the day-to-day difficulties inherent in maintaining a broad-gauge network. They had parallel tracks, one broad and one narrow, on either side of a platform. Two survive, at Exeter (1860), constructed of brick, and at Didcot (1863) in timber, moved a short distance from its original site to the Didcot Railway Centre.

Figure 34
Ross-on-Wye is one of only a handful of Brunel's broad-gauge goods sheds to survive. Built in 1855 of sandstone, it displays the characteristic timber boarding used by Brunel on the end elevations of his later goods sheds.
[DP181264]

Brunel's designs of the 1840s set the pattern for GWR goods sheds for the next 20 years. Sheds continued to be of the through roadway type into the 1870s and timber was still favoured for end walls or at least the portion over the rail line.

By the 1880s, the Brunel influence was finally thrown off with the arrival of much more modern designs having more in common with contemporary industrial buildings, with extensive use of hard engineering blue brick dressings for plinths, cills etc, and segmental-headed iron-framed windows. A distinctive feature was the use of extensive timber-framed glazing, filling almost all the upper part of the gable of one or, in many cases, both end walls. Examples included Slough, Tetbury (1889) (Fig 35), Brent (1893) and Ledbury.

A new and more austere design appeared in the late 1890s. It retained the iron windows and general characteristics of its predecessor but the glazing in the gables was gone, to be replaced by three iron-framed windows with the central one raised slightly above the two that flank it. Examples were at

Figure 35
Tetbury (1889). The distinguishing feature of the late 1880s GWR designs was the full-width glazing of the gables. The increasing using of iron as a structural element is also evident with I-beams over openings and cantilevered brackets for the canopy over the cart loading bay.
[DP173409]

Westbury, Castle Cary, Redruth (1912), Market Drayton, Chipping Sodbury (1903) and Tettenhall (1913). Survivors of the type include a number on the Gloucester & Warwickshire line at Broadway (1904) (Fig 36 and *see* Fig 18e), Gotherington (1906), Toddington (1906) and Winchcombe (1905). By this time small lock-up goods sheds were increasingly being used. For the new shorter route to South Wales opened in 1903, hipped-roof brick structures using much of the architectural vocabulary of their larger siblings were placed on platforms. As with several other railways, economy was to be seen in the increased use of corrugated iron on steel frames, both for new goods sheds such as that at Shipston-on-Stour and for extensions to existing ones. Previously, extensions had been carefully carried out in matching materials; now work was entirely utilitarian.

Major companies absorbed by the GWR, such as the B&ER, had their own standard designs. These ranged from plain brick structures with narrow

Figure 36
Broadway (1904) represents the final development of the traditional goods shed on the GWR. Extensive use of engineering blue brick and hard red brick is apparent. The building, which has been little altered externally, now forms part of a caravan park.
[DP173369]

windows at Hele & Bradnich and Wellington to a handsome group of stone goods sheds on the Cheddar Valley line with window frames of Gothic or Florentine inspiration. A keynote of many B&ER buildings was the use of elaborate bargeboards, and some of the goods sheds, such as Axbridge (1869) (Fig 37), shared this feature. The B&ER sheds differed in their layout from those of the GWR, eschewing the through cart road for a more conventional side doorway. As well as standardisation on the part of companies, there were designs that reflected the work of a particular engineer. Ten independent companies subsequently absorbed or leased by the GWR that had built branch lines in the 1880s had William Clarke as their Civil Engineer.[14] Clarke had a distinctive style of station building, and several goods shed designs that could be executed either in stone or brick. A number of these survive: Avonwick, Gara Bridge and Loddiswell on the Kingsbridge branch (opened 1893) among them.

Figure 37
Photographed in the 1960s, Axbridge (1869) displays the B&ER's enthusiasm for decorative bargeboards, used by the company on stations, signal boxes and goods sheds. Equally characteristic are the finely coursed masonry and the pointed arches.
[R S Carpenter Collection]

London & North Western Railway

Standardisation emerged somewhat later on the LNWR than on some other company lines. A key feature was the widespread employment of bricks made by the LNWR itself, including extensive use of blue brick and panelled brick construction (Fig 38 and *see* Fig 18c). Lighting of the smaller sheds was invariably by roof lights. Most distinctive were the three openings in the gable ends with louvres to ventilate the interior, which are found in examples ranging from small sheds in country goods yards to large multistorey warehouses from the 1860s to 1890s. A typical example is the now demolished shed at Market Harborough (Fig 39). The layout was the conventional arrangement with side openings for carts, often sheltered by an awning formed by an extension of the roof, necessary because LNWR sheds generally had carts loading outside the building rather than cut-outs in the platform to enable them to be backed into the shed.

At the beginning of the 20th century, the LNWR also favoured the trackside type of shed for smaller yards, again with the roof extended to cover

Figure 38
Stafford (c 1880) is an example of a large LNWR goods shed and exhibits that company's preference for rooflights rather than windows in the side walls. The office is a later addition.
[DP173398]

not just the cart loading area but the space between the track and the building. Probably motivated by a need for economy, evident in other companies' designs in the same period, a range of timber goods sheds was developed by the LNWR. These were substantial structures with the track passing through, with an external goods office, an awning over the cart entrance and clad with clapboarding. Like their brick counterparts, lighting of the shed was by a central roof light. As with all timber sheds, the vast majority have been demolished, but a well-preserved example remains at Waverton (1898).

Figure 39
The closest that the LNWR came to a standard design, Market Harborough (demolished) has the distinctive triple-louvred openings in the end gable associated with the company.
[John Minnis Collection]

Midland Railway

The goods sheds built for the line from Leicester to Hitchin in 1857 marked the start of MR standardisation. The earliest examples, of which the sheds at Sharnbrook, Oakley and Wellingborough (*see* Fig 1) survive, set the style with pointed arch blind arcading on the side elevation, similarly pointed window and door openings, and elaborate iron window frames incorporating diamond and lozenge motifs. The 1857 examples had vivid polychrome brickwork in red with cream bands and dressings. All these architectural features formed part of a corporate style and reflected those found in adjacent station buildings. The basic form, albeit with many detail differences, remained in use for almost 20 years and could be found in both brick and stone versions. Examples were built along the Bath branch (that at Bitton survives) and the Mansfield–Shirebrook lines, opened in 1869 and 1875 respectively. The best-known examples were the handsomely executed sheds of 1876 on the Settle & Carlisle line, many of which survive (*see* Figs 18a and 19).

Larger MR goods facilities could be impressive. The depot opened at Worcester in 1868 (Figs 40 and 41) had a succession of gables with blind

Figure 40
The MR opened a goods depot at Worcester, a little to the south of the GWR passenger station, in 1868. As with much of the MR's infrastructure, the work was carried out on a grand scale. The resulting building made use of polychromy and the MR's highly distinctive iron window frames to create an impressive ensemble. [DP 173424]

Figure 41
The office block at Worcester.
[DP173420]

arcading alternating with windows in its side elevations, and a two-storey office. Internally, light iron roof trusses were employed while iron was also used to create the highly distinctive pattern of window frames favoured by the MR in this period. The extensive use of these, combined with cream brickwork to provide relief on gables and on lintels, produced a striking effect.

By the beginning of the 20th century, economy was the watchword and many depots were constructed of timber, usually board-and-batten, and treated with creosote rather than painted.

Lancashire & Yorkshire Railway

For many years, the L&YR built its goods sheds of rock-faced stone, which gave them an appearance of great solidity, very much like early canal warehouses (*see* Fig 20). Such warehouses were also associated with the East Lancashire Railway, which amalgamated with the L&YR in 1859, and a number remain:

Baxenden (1848), Bury (1846), Summerseat (1846) and Rawtenstall (1846). The buildings were often of considerable size, reflecting the company's role in moving so many of the products of its heartland in industrial Lancashire and Yorkshire. The need for warehousing frequently led to the buildings being constructed with one or two additional floors, such as the shed at Waterfoot (*see* Fig 24). Standard designs began to emerge in the 1870s and the sheds at Chatburn (Fig 42) and Clayton West (both 1879) are extant examples of the substantial stone structures of the period. By the 1890s a harsh red brick design was prominent, again often large with warehouse space. Horbury & Ossett, built in 1902, is an example of one of these large sheds, of which a number were put up in conjunction with line widenings at the time. The cotton shed at New Hey (1913) employs the same elements of large, rectangular iron windows, panels and heavily corbelled gables.

Figure 42
Chatburn (1879) typifies the solid structures erected by the L&YR in the 19th century. The L&YR sometimes favoured goods offices that were visually separate from the main body of the goods shed, instead of the more common arrangement of a lean-to or wing. [DP185885]

Furness Railway

Even small railway companies could have elaborate standard structure designs. Among the most impressive was the FR, which retained Paley & Austin as architects. Documentary evidence as to exactly which buildings they were responsible for is often lacking, but the standard FR goods shed, built from the late 1860s to the 1880s, can probably be attributed to them. This could be of brick, which had vivid polychromy, as at Haverthwaite, or stone, with prominent buttresses, as at Seascale or Grange-over-Sands (Fig 43). Common to both was the employment of large lunettes.

Figure 43
Grange-over-Sands. The Furness Railway's standard design was another that employed lunettes but was made especially distinctive by the use of heavy buttresses to the piers that divided the buildings into bays. The infilling of the opening for the rail entrance is typical of the unsympathetic alterations made to many goods sheds.
[DP169037]

Great Eastern Railway

The Eastern Counties Railway, the principal constituent of the GER, favoured clapboarded timber goods sheds and many examples of these, built in the 1840s, survived well into the post-war period. Even such major traffic centres as Bury St Edmunds, Stowmarket and Thetford had timber sheds, the former lasting until 1954 when it was in a state of considerable disrepair (Fig 44). Brick sheds did exist, and those at Attleborough and Wymondham (Fig 45)

Figure 44
Bury St Edmunds. The sheer decrepitude into which some goods facilities had fallen by the 1950s is demonstrated by the 1840s shed at Bury St Edmunds, which had been condemned in 1913 but is seen here on 27 March 1952. Its timber construction was typical of many early sheds built by the GER's constituents. Its replacement is shown in Fig 69.
[British Railways, courtesy D Middleton]

Figure 45
Wymondham (1845), a Norwich & Brandon Railway structure with four-centred arches and, like many early goods sheds, built at right angles to the running lines, with access via a turntable.
[DP172462]

(both 1845) on the Norwich and Brandon Railway were essays in Tudor Gothic, taking their cues from the company's station buildings. But these, like other goods sheds of the same period, were really line styles. A push towards standardisation came following the formation of the GER in 1862 with what has become known as the '1865 type' of Italianate station buildings, used on a number of new lines, and which had an equivalent goods shed design. One example remains at Clare (1865). But, other than the '1865 type', there is little evidence of standardisation, with line styles being the predominant approach.

Figure 46 (left)
Tattershall (1848) is one of a number of GNR sheds in Lincolnshire. Key features are the lunettes and the kneelers on the gable edges, with the goods office a later addition. Externally, the shed is exceptionally well preserved.
[DP172147]

Figure 47 (below)
Little Bytham was built in 1912 as part of the main line quadrupling carried out in that year. The roof extends to form a continuous canopy over the cart loading area and a large window in the gable provides light for the interior. Photographed in 1998, the building was subsequently demolished.
[John Minnis]

Great Northern Railway

The GNR initially favoured a handsome goods shed design of the parallel cart road type with gabled ends, terminating in kneelers, lunettes, and twin round-headed openings at each end. Examples are to be found at Bardney and Tattershall (1848) (Fig 46). The design was remarkably similar to one used by the South Eastern Railway at the same time at Wateringbury (1845). Much later, around 1900, a distinctive design was used where stations had been rebuilt for line-widening schemes. This had a circular window in the gable and was generally constructed of light-coloured Peterborough bricks, with red headers providing plenty of contrast. A well-known example is that at Little Bytham (1912) on the East Coast Main Line (Fig 47).

North Eastern Railway

A multiplicity of styles and designs was to be found on the NER, reflecting its complex origins. The lines associated with George Hudson – the Y&NMR, the GNER, etc – rivalled those associated with Brunel for the early application of standardised designs. For those companies, G T Andrews built a number of goods sheds, of which 11 survive in various states of preservation at Bridlington (1846), Cottingham (1846), Ganton, Hutton Cranswick (1846) (*see* Fig 16), Leeming Bar (1848), Nafferton (1846), Pocklington (1847), Scarborough (1845), Stamford Bridge (1847), Thorp Arch (1847) and Wetherby (1847).[15] The quality of Andrews's designs is recognised in the high proportion of survivors that are listed. The goods sheds are notably large in comparison with those of other contemporary railways, varying in length but always 36ft in width. The initial group of sheds were of the through type with a cut-out in the platform to accommodate carts. The sheds on the Harrogate–Church Fenton and York–Market Weighton lines had a through cart road running from end to end with no cut-outs in the platform, although there were still side entrances as well as those in the ends. A distinctive aspect of the design was that, in many examples, the hipped roof also covered an open-sided loading bay at one end of the building, supported on a double row of five hexagonal cast-iron columns. Many were subsequently removed but those at Pocklington, Nafferton and Thorp Arch are retained, although the first two are now enclosed. Andrews also used the same cast-iron columns as the basis of smaller open-sided goods sheds on the Scarborough line of 1845, the last survivor of which, at Seamer, was demolished in 1978.

Another noted railway architect, Benjamin Green, was responsible for an extensive series of standardised goods sheds on the York Newcastle & Berwick Railway. Again, they reflected the architecture of the stations at which they were located – in this case, Tudor Gothic. Constructed in finely executed masonry, they had stone balls topping each steeply pitched gable, matching those on the station buildings, and arched openings, hoodmoulds and substantial buttresses. Two survive at Acklington (Fig 48) and Christon Bank, both of 1847.

Figure 48
One of Benjamin Green's Gothic designs for the York Newcastle & Berwick Railway, Acklington (1847) is seen here disused in August 1978. It has subsequently been converted to a house.
[BB81_05015]

Successive NER company architects produced their own standard designs, all of them through type with cart bays set into the platform. In 1856, a hipped-roof design was employed, examples of which continued to be built for some years, with Ripon (1871) a late example. In 1860 Thomas Prosser designed a somewhat simpler gable-ended shed which was built in quantity as a system-wide standard both in brick and in stone over the next 20 years. Details included the extension of the roof over the cart entrance, flat-headed iron windows and sliding doors under cast-iron lintels. Hexham (1873) (Fig 49) is an example of one that is double the usual length. As with other companies, the NER developed a smaller shed of the trackside type. Sometimes these were constructed with an awning over the cart entrance as at Riding Mill (1877, demolished), sometimes without the awning as at Glanton (1887) on the Cornhill branch. Timber sheds were employed in areas where subsidence was a problem and the same material was also used for platform lock-up sheds, as at Levisham.

Figure 49
An NER standard design by its company architect, Thomas Prosser, at Hexham (1873), as converted for use as a wine retailer. This example is, in effect, two of these sheds joined end on.
[DP169028]

5

Large goods sheds and warehouses

While the vast majority of goods sheds were relatively simple structures, accommodating just a few wagons each, cities and industrial towns were served by much larger buildings, many with warehouse space. In addition, specialised types of warehousing were constructed, for example those for storing food such as grain, potatoes and vegetables and for textiles such as cotton and wool, together with even more specialised uses such as gunpowder.

Many urban goods sheds were not radically different from their country counterparts, just much larger in size. Cart entries in side walls were more numerous, with up to six being common, platforms were wider and methods of handling the goods were more sophisticated than the 30cwt jib cranes found in smaller examples. It was the scale that was different rather than the basic principles of operation. An early surviving example is the 1849 L&YR warehouse at Halifax Shaw Syke (Fig 50), which has timber floors supported on cast-iron columns, with both of the long elevations curved. One common design feature was a multi-pile roof structure so that a major goods shed such as that at Camden (1864) (Fig 51) resembled half a dozen smaller sheds lined up side-by-side. The larger the shed, the more important it was that it was adequately lighted, and large sheds tended to have serried ranks of iron-framed windows and extensive roof glazing to illuminate the interior. Major goods depots were often located in manufacturing districts, and the steam from the engines in the yards combined with the generally smoky atmosphere to produce conditions that could be taxing. Another common feature was the addition of a second floor for warehousing. Two-storey sheds were often distinguished, from the 1890s onwards, by the extensive use of structural steelwork internally to support the considerable weight of goods stored on the upper floor. Bedford, MR (Fig 52) is an example where massive joists of steel are exposed. Access and egress from the upper floor could be achieved internally by trapdoors in the floor through which goods could be hoisted up, or externally through taking-in doors or lucarnes, as was the case at Bedford. Upper floors were typically used for the storage of grain, while some warehouses also had cellars which could be used for storing ale or meat.

Stylistically, such goods sheds had much in common with their smaller counterparts. There was the same use of panelling to divide a lengthy building into a series of bays, and architectural features such as the triple vents on the gable ends of LNWR sheds (*see* Fig 39), the ornamental iron windows of the MR (*see* Fig 18b) or the semicircular fanlights of the LB&SCR (*see* Fig 18a) were repeated.

The goods yard at Leicester's MR station is an example of how individual companies had their own facilities in a town. The large shed to the left and the four-storey grain store above it served the MR, while the LNWR opened the goods shed to the right of them in 1875. It was joined by a much larger combined LNWR goods shed and multistorey warehouse in 1898 (the sole survivor today), seen on the far right. Some ancillary structures are also visible: an accumulator tower and power house just to the right of the MR grain warehouse and two sets of stables, one next to St Luke's church and the other just below the 1898 LNWR warehouse. Just visible at the top right-hand corner are the goods sheds and warehouse of a third company, the GNR, at Belgrave Road, opened in 1882. The circular structure in the foreground is the locomotive roundhouse, which was newly completed when this photograph was taken on 27 June 1949.
[EAW024245]

Figure 50 (facing page)
At Halifax, goods traffic was shared between the L&YR, the GNR and the LNWR. The earliest goods shed visible in this view of September 1931 is the curving structure at the bottom, the L&YR warehouse of 1849. Just above this is the open-sided fruit shed of the L&YR, still extant albeit derelict and above it to the right, the distinctive 1885 L&YR/GNR goods warehouse with its curving east wall. [EPW036888]

Figure 51 (above)
Camden's goods shed of 1864 with its eight gables is in the centre of this photograph of 16 September 1921. To the right is the still extant warehouse of 1905, with direct access for boats from the Regent's Canal: the entrance can just be made out under the towpath bridge. [EPW007016]

Figure 52 (right)
A substantial goods shed with warehousing on the first floor was built by the MR at Bedford in the 1890s. Built of the hard red brick employed by the MR, the shed made much use of structural steel in its construction but still retained traditional decorative elements such as the eaves corbelling and bands. Access to the upper floor is via hoists in lucarnes, and extensive clerical accommodation was provided by a two-storey goods office. The building remains in railway use as Network Rail offices. [John Minnis/Historic England]

As more staff were required to carry out the complex paperwork involved in goods traffic, goods offices tended to be much larger and were often of two storeys. Unlike with small sheds, the owning railway companies often took the opportunity to advertise their services on their buildings, on large signboards attached to the building, by lettering painted directly on the brickwork or by the company's name spelt out in light-coloured bricks high up on the façade. The publicity-minded L&YR was a great proponent of this approach. Some instances of it are at the massive warehouse at Bolton, demolished in the 1980s, at the Fleetwood grain elevator (also demolished), where the directors voted to spend an extra £60 to display the company's name in white glazed bricks, and at the extant New Hey cotton warehouse. Painted signage survives at the MR warehouses at Burton upon Trent (Fig 53 and *see* Fig 66) and on the GNR

Figure 53
*Reused as a Travelodge hotel, the MR No 2 Grain Warehouse (built as a cheese and corn store), adjacent to Burton upon Trent station, has retained its lucarnes. Burton once had many structures of this type but only this and the bonded stores (*see *Fig 66) remain, both listed Grade II. The grain warehouse is a typical example of these specialised structures, with small windows on the two upper floors and large round-headed windows on the ground floor. The conversion has retained the basic form of the building although there has been extensive renewal of windows. The painted white lettering on a black ground is also characteristic of this type of warehouse. [DP172901]*

goods shed at Morley. The most impressive surviving example is the CLC warehouse at Warrington, which has concrete panels displaying its name and those of the three companies that made it up (Fig 54).

A special type of large goods shed was the transhipment shed. Although this term has subsequently been widely used in describing goods sheds (which all involve the transhipment of goods from one mode of transport – the horse cart or lorry – to another, the railway wagon), in the 19th century it was used in a very specific way by railway companies. It referred to a shed where goods, instead of being brought by local merchants for transport by rail or deposited for them to collect or be delivered, were transferred from one wagon to another as part of a lengthy journey across the country, with part loads being placed together in wagons to ensure that when the train left it comprised full loads

Figure 54
The Cheshire Lines Committee warehouse at Warrington (1897) made full use of the advertising possibilities offered on its façade to promote the three companies that went to make up the CLC: the GCR, the GNR and the MR. The building, which is now converted to apartments, is an example of an ostensibly utilitarian structure, built relatively late, which still made extensive use of ornament. The proportions were also carefully handled, with effective repetition of elements along the lengthy symmetrical façade and the breaking of the advertisement panels by openings for hoists (now converted to windows).
[DP169039]

to the various destinations. Thus, a transhipment or tranship shed, while it externally resembled a conventional goods shed, had an entirely different purpose. The most famous example was at Crewe LNWR (demolished) but there was an increasing move towards this type of operation around the beginning of the 20th century. A second meaning given to the term in the 19th and early 20th centuries was the transhipment of rail to water traffic and vice versa.

Other features found in larger yards were the ancillary buildings needed to service such facilities. Chief among these was extensive stabling. The 11 largest railway companies had some 26,000 horses in 1914 and all major goods yards had ranges of stables, which are a subject in themselves. Even relatively small yards might well have a two- or three-horse stable and, as with goods sheds, railway companies had their own distinctive designs for these. Stables were usually located within the goods station complex but often on the edges of it so as to cause the minimum of distress to the horses. The large groups could often accommodate several hundred horses, usually in a single-storey layout (Fig 55), but in confined areas where property costs were particularly high, stables could

Figure 55
Stables at Queen's Road depot, Sheffield, MR (1892), typical of those found in larger goods yards throughout Britain. Photographed on 7 May 1962.
[D Middleton]

be multistorey with lengthy ramps leading up to upper levels, for example at Paddington GWR and Willow Walk LB&SCR. The best-known survivors are probably those at Camden (built in several phases from the mid-1850s, listed Grade II) but there are a number of other surviving examples including stables for 600 horses at Paddington GWR (1876–83) and two large and little-known GER complexes at Hare Street (1881) and Quaker Street (1887) outside Bishopsgate that once served the warehouses both on the Bishopsgate site itself and adjoining the GER main line. In addition to the stables themselves, there would be facilities for vets and sometimes – as at Camden and King's Cross GNR – a horse hospital and a house for the head horseman. By the 1930s, many of the stables were being converted to garages for the increasing numbers of lorries employed by the railways, but as late as 1947 the railways still had over 9,000 horses at work on collections and deliveries in urban areas.

Another ancillary requirement concerned the way in which wagons were moved, both within the yard and inside the goods sheds. Steam locomotives were not allowed to enter goods sheds, many of which had a notice to this effect by the entrance. Wagons were moved in one of three ways: by hand, by horse or by capstan. The first two were by far the most common, especially in smaller and older yards, but the introduction of hydraulic power from the 1860s made large multistorey warehouses much more practical. The hydraulic plant was located in an accumulator tower with a water tank nearby. Several such towers remain in locations where the goods warehouses they served have been demolished. They include a number of examples in Poplar where an MR boiler house, engine house and accumulator tower of 1881–2, together with a pair of NLR accumulator towers of 1877 (listed Grade II), survive. A boiler house, engine house and accumulator tower of 1886 (listed Grade II) that served the LT&SR Commercial Road depot is a further East End survivor. Others survive outside London, for example at Leicester (MR), Burton upon Trent (MR) and Huddersfield (LNWR) (Fig 56). The towers are distinctive structures, the MR favouring a tall, hipped-roof, brick-built structure with three louvred round-headed openings near the top, giving the tower an Italianate character. From 1900, electricity started to supersede hydraulic power but many hydraulic installations remained in use until large goods depots of this type were phased out in the 1960s.

Capstans are especially associated with the use of turntables (Fig 57). These are central to understanding how large urban yards and warehouses

Figure 56 (facing page)
The boiler and pump house and accumulator tower at Huddersfield, supplying power to the LNWR warehouse.
[DP169343]

Figure 57 (right)
The interior of Curzon Street goods depot, Birmingham, photographed in 1966 shortly before demolition, shows the turntables and capstans and the insetting of the rails in stone setts, all frequently to be found in large warehouses.
[BB64/021]

worked, and the job of capstanman was a highly skilled one. Wagons could be easily directed to where they were required by means of these turntables, which saved space in confined urban surroundings. They were to be found in the yard itself and within the buildings, whose platforms had rounded cut-outs to accommodate them. Wagons could literally be manoeuvred around the building by means of ropes attached to the wagons and moved by the capstans via the turntables. The use of these turntables is significant: their small size, along with other factors such as cramped sidings within industrial sites and coal mines, restricted the development of the British railway wagon. Unlike the wagons in mainland European railways, these changed little in size from the mid-19th century until the 1970s, one factor that helped destroy the long-term viability of moving goods by rail.

This brings us on to a discussion of the largest type of building catering for goods movement by rail, the multistorey warehouse. As we have seen, the principle goes right back to the earliest days of railways, with the Manchester, Liverpool Road warehouse of 1829 being the prototype of all that followed.

Many were built, few survive. After Manchester, Liverpool Road, the MS&LR 'London Warehouse' at Ducie Street, Manchester (1867) (Fig 58) is the next-oldest survivor of these large warehouses. This was of seven storeys, its massiveness accentuated by small, almost square windows. The general layout for these structures had a goods shed at ground floor level where wagons would enter for loading and unloading. Goods not collected by the customer or

Figure 58
The MS&LR's Ducie Street, Manchester goods station is seen in a photograph taken in June 1932 to record a nearby warehouse fire. Immediately in front of the source of the fire is the multistorey warehouse, all that exists today of the complex, with goods sheds to the right. On the extreme right with prominent lucarnes is the MS&LR grain warehouse and on the left, adjacent to the entrance ramp to Manchester London Road station, is the triple-pile LNWR goods warehouse.
[EPW038687]

delivered by cart by the railway would be stored on the upper warehouse floors on behalf of the customer, at a charge. As time went on, with the introduction of hydraulic power it became possible to provide hoists (Fig 59) that could move wagons to upper floors or in some cases basements of the warehouse, where they could be moved by the aforementioned turntables and capstans. In some cases, such as the LNWR Manchester, Grape Street warehouse (1869),

Figure 59
The Huddersfield LNWR/L&YR warehouse (1885) with its external, hydraulically powered wagon hoist for taking wagons to the second floor of the building. [DP169344]

part of the Liverpool Road complex, the building could be accessed by rail from a viaduct rather than at ground-floor level. Conditions in some of the depots built on confined urban sites could be claustrophobic. At Bishopsgate (*see* Fig 13), the basement level of the goods depot contained short sidings built within the arches of the original Eastern Counties Railway viaduct; conditions within the even more confined spaces of the GWR's Smithfield depot, handling the meat traffic of the market, must have been trying.

Typically, such buildings would have a number of tracks running through them, up to 10 in some cases, with some linked internally by turntables. The design of the buildings altered little for many years and, despite the value of the goods stored in them, fireproofing lagged behind that in many other industrial buildings, with floors and internal platforms often being of timber. Fires were frequent: the L&YR warehouse at Huddersfield burnt down in 1867 and its replacement of the following year was also destroyed by fire only a few months after it opened. A third warehouse on the same site, built in 1869 (Fig 60),

Figure 60
The 1869 L&YR warehouse at Huddersfield.
[DP169348]

remains to this day, although it still had wooden floors and roof trusses. However, by the late 1860s, the increased use of iron was evident. The London warehouse at Ducie Street, Manchester had an internal structure of cast-iron columns, riveted iron box girders and brick jack arches. The LNWR/L&YR warehouse at Huddersfield (1885) (*see* Fig 56) was of similar construction at first-floor level, although the floors above were still of timber on iron beams and columns. It was also somewhat more decorative in appearance than Ducie Street, with a dentilled cornice and much larger windows. The LNWR continued to build similar structures well into the 1890s; the one at Leicester, built in 1898, is a notable survivor (Fig 61 and *see* Chapter 5 opener). A strong industrial aesthetic allied to their sheer scale ensured that these urban warehouses were impressive structures having much of the visual impact of contemporary mills. For example, at Halifax (*see* Fig 50), the L&YR/GNR warehouse of 1885 had bays topped by individual gables and was enlivened by

Figure 61
The multistorey LNWR warehouse and goods shed at Leicester, built in 1898. It had ornament provided by the cornice and bands of straw-coloured brick to provide relief to the pervading bright red bricks used throughout the exterior. The land rises to the east so that the structure is actually two storeys higher on the road side. The tallest part of the structure has rail access at third-floor level, with two floors below that and two floors of warehouse space above it. Also evident are the high boundary walls to be seen in such locations.
[DP172910]

pilaster strips and corbel tables. Its eastern flank followed the curve of the tracks so that the entire façade was gently curved.

The development of traditional warehouses reached its zenith at the turn of the 20th century. At Marylebone, the GCR had a vast structure that provided two acres of storage on each of the three warehouse floors (Fig 62). The goods station was located on the ground floor and a basement provided additional warehouse space. The interiors were lit by nine shafts piercing the building

Figure 62
Marylebone GCR goods yard in July 1923. The warehouse, with its covered loading platforms, is the most prominent structure, with the goods offices on the extreme right. To the left is the power station for the depot and, below that, the lengthy curving roof of the transhipment platform for interchange with the Regent's Canal.
[EPW009006]

through the three upper floors, through which goods were hoisted by overhanging hydraulic cranes.[16] The five-storey GNR warehouse at Manchester, Deansgate (Fig 63 and *see* Fig 3), designed by the engineer W T Foxlee, was the most spectacular of all. Completed in 1899 and 267ft long by 217ft wide, it was of fireproof construction, making much use of structural steel along with riveted wrought-iron girders, within a brick outer shell. The design was innovative inasmuch as, to make the best use of the restricted space open to a

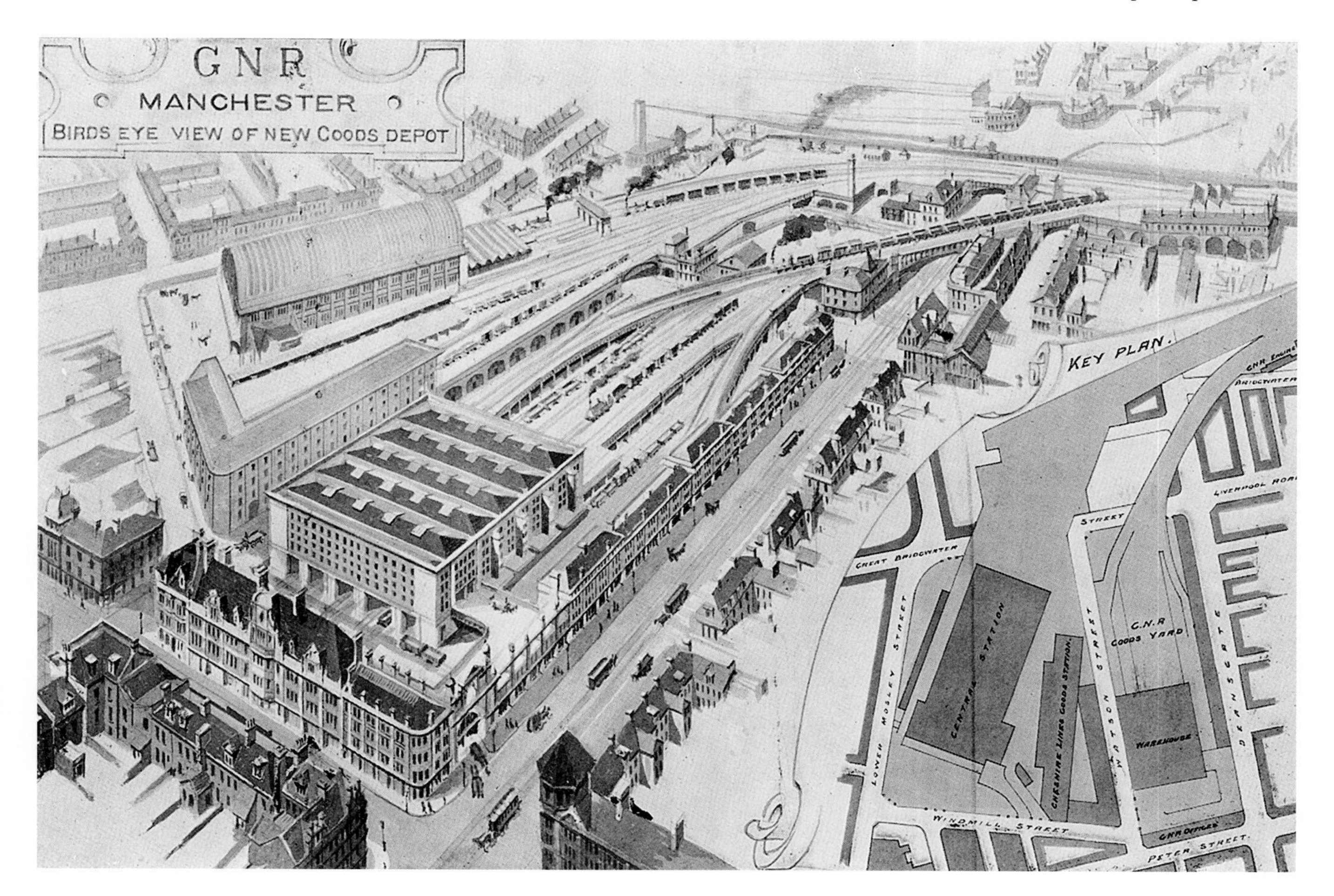

Figure 63
The GNR prepared a perspective drawing, reproduced in The Railway Engineer, *to publicise its Manchester, Deansgate, warehouse.*
[From C H Grinling, The Ways of Our Railways, *1905]*

latecomer in the heart of a city, two separate goods stations were constructed, one on top of the other, with distinct high- and low-level yards linked by an incline. Above them were three warehouse floors. It was common for large warehouses of this type to fully use the topography of the site, sometimes entered at more than one level and with wagons transported internally by hydraulic lift, but to have completely separate goods stations within the same structure, the upper one for London traffic and the lower one for other destinations, and each with its own yard, was entirely new.

Office accommodation for large yards was often within a separate building, frequently of considerable size and architectural pretension, for example the five-storey goods offices at Paddington of *c* 1906 that also housed the GWR's Chief Goods Manager and his staff (Fig 64). A surviving example is at Nottingham (1875, listed Grade II), where the office building is the only remaining part of a once extensive goods station. Such buildings were often located on the edge of the yard, facing the public road, providing a point of contact with customers, while the remainder of the yard was hidden behind high brick walls with but one point of entry, controlled by a gatekeeper's building, to discourage pilferage. At Marylebone GCR (*see* Fig 62), the office building was a substantial three-storey building with a large opening on the ground floor forming the entrance to the goods yard.

Specialised warehouses were most commonly found within large complexes, although they could be present in smaller yards, as at New Hey, L&YR, where a large cotton warehouse was built in 1913. A typical example of a large yard was Sheffield, Bridgehouses, MS&LR, which had, besides a large goods shed, a covered loading bay, a massive multistorey granary and a four-storey potato warehouse, capable of holding some 100,000 bags of potatoes. Covered loading bays for fruit and vegetables were common at major goods depots but are now almost extinct. Halifax retains in derelict state an L&YR fruit shed which is open-sided and has a north-lit roof (*see* Fig 50). At Nottingham, the MR had a two-storey goods and grain warehouse, a seven-storey grain warehouse, a bonded store and a fruit and potato store (all demolished). Granaries were generally large multistorey structures such as the noted example at King's Cross (*see* Chapter 7 opener). Smaller structures were provided at rural locations, although even these, such as the three-storey granary at Lingwood, GER (1884) (Fig 65), were generally larger than most goods sheds. More

Figure 64
Paddington goods depot was rebuilt in 1925–31. A vast steel-framed shed, with corrugated iron cladding and no pretension to architectural effect, was put up. The large multistorey building beyond the structure is the 1907 goods office block, which was retained, as was the lengthy four-storey warehouse to the left of the new structure. The passenger station at Paddington, with its overall roof, is visible on the right in this photograph of 28 July 1938.
[EPW058334]

G W R
PADDINGTON GOODS STATION

Figure 65
A rare surviving rural grain store, Lingwood was constructed in 1882. The heavy corbelling on the gable ends was also to be found on the adjacent passenger station and exemplifies the employment of a house style. It was the work of the GER's architect, Edwin Wendover.
[Mike Page, www.mike-page.co.uk]

Figure 66
The second MR warehouse at Burton upon Trent to survive is the large No 4 Bonded Stores and Grain Warehouse. This is still in warehouse use and consequently is less altered than Grain Warehouse No 2.
[DP172874]

sophisticated forms of storage were rare, with little to rival the enormous grain silos to be found in the American Midwest, although the L&YR built an imposing grain elevator at Fleetwood in 1881 (demolished). Some of the most impressive specialised warehouses were the great bonded warehouses provided for brewers and wine shippers. These were particularly associated with the MR, which built St Pancras above vaults which were used by Bass as its London store. In Burton upon Trent, two large bonded warehouses survive (Fig 66 and *see* Fig 53) and at Derby a grain warehouse and a bonded store form an important group.

6 The 20th-century goods shed and warehouse

The Edwardian period

Large numbers of goods sheds and warehouses were built up to 1914. Smaller goods sheds continued to be constructed in much the same way and to similar plan forms as those in the 1880s and 1890s. They included sheds for new branch lines and replacements of older sheds on existing lines. The most noticeable tendency was towards simplification of external appearance, with decorative touches such as polychromy, brick cogging and ornate iron window frames disappearing. More sheds were built with steel frames clad in corrugated iron and these materials were often used to expand existing sheds, with no attempt to match earlier work as had usually been the case in the 19th century. In the larger examples, considerable use was made of structural steel to support upper floors and even in the smaller structures steel increasingly replaced timber and iron for door lintels. New facilities, some of considerable size, were constructed to serve the burgeoning textile and engineering industries in the north.

Large multistorey warehouses remained in vogue. Initially, these were built on traditional lines but, increasingly, reinforced concrete framing was favoured by some companies. It was introduced into railway buildings by the GWR in 1899 at Brentford Dock (Fig 67). Substantial ranges of four-storey warehouses (demolished) were erected using the Hennebique system, just two years after the first example in Britain of this type of construction was built in Swansea. The system was taken up by W Y Armstrong, the GWR Engineer for New Works, from 1904. His first building of this type was a two-storey grain warehouse at Plymouth Docks (demolished). The next, in 1906, was a two-storey warehouse (designed in 1904 and also built to Hennebique patents) at the Canons Marsh depot, Bristol. The architectural work, by P E Culverhouse, was simple, with plain concrete walls, open on one side on the ground floor and with a flat roof. Internally, two broad parabolic arches spanned the central bays, supported on square-section reinforced concrete columns. The Grade II listed warehouse remains, but the adjoining steel-framed transit shed, clad in corrugated iron and with a light steel truss roof, has been demolished.

Canons Marsh was on a relatively modest scale, but the GWR's South Lambeth depot (Fig 68), opened in 1913, was the company's first really modern depot inasmuch as it had no fixed cranes to obstruct its platforms, with all lifting carried out by one-ton overhead electric cranes and mobile cranes. It was

The goods depot at Bristol Temple Meads, GWR was completely rebuilt in the 1920s, the work being completed in 1929. The depot exemplifies the move in the inter-war years towards large single-storey structures, steel-framed with corrugated iron cladding, often with large open-sided areas. The sheer scale of such facilities, close to a city centre, is clear from this aerial view of 1938. Following closure of the depot, the site has now been redeveloped.
[EPW060120]

Figure 67
The GWR was a pioneer in the use of the Hennebique system of reinforced concrete construction. The company made its first use of it in a warehouse at Brentford Dock in 1899, the structure with the prominent lucarne visible in the background. Further warehouses were constructed at the dock prior to 1914 which employed brick infilling to the clearly expressed concrete frame, as seen in the foreground.
[AA001574]

also on a massive scale, 400ft long, with a reinforced concrete frame clad in brick with much of the framing visible, and had three storeys of warehousing above a ground-floor transit shed. Three tracks ran through the building, served by two platforms and, again, much of the ground floor was open, protected by expansive corrugated iron canopies and by roller shutters, used by the GWR for the first time. It was closed in the 1970s and demolished.

The other pioneer of reinforced concrete construction was the NER, which built its New Bridge Street depot, Newcastle, in the material in 1903. This was a three-storey warehouse with two warehousing floors above the tranship depot, designed in a stripped Classical style. It was bombed in World War II, patched

Figure 68 (facing page)
The GWR established a major goods depot south of the river at Battersea in 1913. Known as South Lambeth, it was the company's first modern depot. The three-storey concrete-framed warehouse had one-ton overhead electric cranes and open-sided loading bays that could be closed with roller shutters. Additional large warehouses of more utilitarian design, added in 1929, can be seen both above and below the original structure in this view of March 1933. The 1913 goods office block, with its hipped roof, fronting onto Battersea Park Road, is visible to the right of the lower warehouse. The LB&SCR's Battersea goods shed (just to the left of the south chimney of Battersea power station) is dwarfed by comparison.
[EPW040850]

up but subsequently demolished in the 1970s. The NER also used reinforced concrete in the three-storey extension to the Forth Goods depot in 1906 (*see* Frontispiece). The extension survives, together with a concrete-framed brick-faced office range, although the main part of the depot, built in 1871–4, has been demolished.

Perhaps the overriding impression, however, is that, despite these advances in warehouse construction, British railway warehouses remained wedded to outdated manual handling methods. There were exceptions. The L&YR was a pioneer in this field with, by 1907, its Bradford wool warehouse equipped with electric cranes on each floor, covering the entire floor area, and electric overhead traversers in its Bolton cotton warehouse. Overall, however, there was a great contrast with the USA, where freight houses were increasingly being constructed with modern handling methods as an integral part of their design. In the USA, electric trucks were in widespread use from around 1915;[17] in Britain, men pushing hand trucks was still a familiar sight in many goods depots into the 1930s and beyond. In 1914, it was argued that, in Great Britain, modern handling methods were largely confined to specialised warehouses.[18]

Between the wars

After 1918, the rate of construction of new goods sheds and warehouses was significantly lower. In part this was due to the small number of new lines being opened, but the tendency was to extend rather than to replace existing structures. The grouping of the railways in 1923 into four major groups, following the passing of the Railways Act 1921, led to the rationalisation of some goods facilities. The growth of road competition led some of the railway companies to start offering long haul by rail and then door-to-door distribution from railheads by road, combining the advantages of both forms of transport. Cutting out manual handling, with the opportunities it provided for damage, loss and delay, became the subject of much research by the railways, and the answer was held to be containers. Often thought of as a recent innovation, containers have been in use for specialised purposes since the mid-19th century. In Lancashire, they were widely used for coal traffic and elsewhere principally for furniture removals, while the L&YR had employed 'flats', a type of container, for its cotton goods traffic. In 1926–8, the LMS constructed nearly 500 containers. Numbers continued to increase throughout the 1930s, although they were but a small fraction of the total number of wagons employed, and the containers, no more than 15ft long, were a far cry from those of today.

A 1941 handbook of modern railway operation set out some ideals in the design of large goods sheds and warehouses.[19] A factor of central importance was clear unobstructed spaces within the building. Consequently, columns were to be avoided wherever possible, as were fixed jib cranes, to be replaced with overhead electric cranes. Turntables were another anachronism. As the handbook points out, the confined nature of many depots meant that the railways were slow to take up moving belts to speed the passage of goods and eliminate much of the manual handling within the depot, because of lack of space for the installation.

As with the period prior to 1914, the GWR was in the forefront of new construction, reconstructing many of its major depots between the wars. These included Paddington (*see* Fig 64) and Bristol (both completed 1929), Swindon (1932), Bordesley, Birmingham (1931), Soho & Winson Green (1933), Cardiff and Swansea, together with a more than doubling of the size of South Lambeth in 1929–32. Of all these, only the Swindon goods shed and the warehouse at Bordesley survive. Bordesley had three warehouse storeys above the ground-floor tranship shed and was a plain concrete-framed building. Access to the upper floors was by electric lift. Stylistically, it had much in common with the other warehouses built by the GWR from 1929 onwards, with an austere appearance determined by the use of small-paned, steel-framed windows set within the externally expressed concrete frame. Reinforced concrete was favoured only for multistorey warehouses: large goods sheds were steel-framed, with corrugated iron cladding and open sides. The lack of any architectural embellishment whatsoever contrasted strongly with the 19th-century structures, most of which did make an attempt, at least to some extent, to appeal to the eye of the traveller or resident.

The SR, too, favoured corrugated iron for its goods sheds. They were mostly put up in conjunction with major station rebuildings, including those at Margate, Hastings and Haywards Heath. Of these, only Margate survives. Again they reflect the lack of concern with appearance as far as goods traffic was concerned. The SR invested considerably in creating impressive new stations, including the magnificent Ramsgate (1926), designed by a young Maxwell Fry. But the goods sheds were absolutely utilitarian structures with little, other than their size, to make them stand out.

The LMS carried out very little reconstruction work, instead concentrating its energies on making its existing depots work more efficiently.[20] It was a pioneer in the use of job analysis and research and introduced electric trolleys, belts and conveyors, although their use was not widespread. It rebuilt Lawley Street, Birmingham, following the destruction by fire of the MR goods shed and warehouse there in 1937. The replacement shed of 1944–5 was again a vast single-storey structure in unpainted corrugated iron. It was intended to be the prototype for further rebuilding schemes but, in the event, post-war austerity put a stop to any concerted programme of reconstruction.

The LNER was the most cash-strapped of the four companies. It carried out little rebuilding of goods warehouses, instead devoting its energies to new marshalling yards, such as that at March, or to places where economic activity was expanding, such as Scunthorpe, where a new goods shed was built. Only where it was absolutely essential was rebuilding undertaken, such as the replacement of the decrepit 1840s timber shed at Stowmarket with a concrete and brick structure in 1926.

Post-war changes

World War II wrought tremendous destruction of goods facilities, particularly those in London, among them the warehouses at Marylebone, Royal Mint Street near Fenchurch Street, and Spitalfields. Reconstruction in many cases consisted simply of patching up the bomb-damaged structures, such as the GNR Farringdon Street depot and Newcastle, Great Bridge Street, which staggered on into the 1960s. Otherwise, rebuilding was confined to the very worst examples, such as at Bury St Edmunds, where an old timber shed that had been condemned in 1913 was replaced in 1954 with a concrete-framed structure (Fig 69). New work in the 1950s consisted mainly of the extensive provision of small prefabricated concrete warehouses for seed merchants at rural stations.

By the time funds were available for reconstruction, it had become clear that cramped city centre sites were no longer suitable for the needs of a modernised railway, and the opportunity was taken to close many old small depots and replace them with large ones on less restricted sites. Such proposals were contained in the 1955 British Railways Modernisation Plan, which set out the future of freight services and envisaged a £385 million expenditure on them, with a reduction in the number of depots and the total or partial closure of 150 marshalling yards.

A significant instance of this was at Peterborough, where several depots were closed and replaced by a new sundries depot on the north side of the city in 1960. This was the one example of a depot with architectural ambition built during this period. The 572ft-long outwards side had a striking concrete barrel roof which contrasted with the flat reinforced slab roof on prestressed concrete beams on the inwards shed. The depot could handle 301 wagons simultaneously with 116 under cover. The wagons were positioned by electric capstans and goods were taken from the wagons by a 331ft-long slatted conveyor to bays where road vehicles were parked. The building was light and airy by comparison with past depots, with extensive glazing in the roof and in the windshields at the north end.

Other examples of the concentration of goods facilities were to be found at Stockton-on-Tees, where a new 60,000 sq ft depot for outwards traffic was opened in 1961; at Gateshead, where a new Tyneside sundries depot was built

Figure 69
The new goods shed at Bury St Edmunds, photographed on 27 August 1953, soon after completion. The concrete-framed structure with extensive glazing is a far cry from its decrepit predecessor (see *Fig 44*)*, although the open side for vehicular access and large end opening must have made it a draughty place to work.*
[British Railways courtesy D Middleton]

on the site of a former locomotive shed in 1963; and at Sheffield, where the Grimesthorpe sundries depot was opened in 1965 as part of a major scheme that involved the closure of four depots spread around the city centre and the construction of the state-of-the-art Tinsley Marshalling Yard. Tyneside and Grimesthorpe were low, steel-framed sheds with corrugated metal siding, largely indistinguishable from other warehousing and light industrial buildings being put up at the same time, and were indeed the predecessors of the 'big sheds' used for warehousing and distribution today. New depots went up at Hither Green for continental traffic in 1960 (Fig 70) and at Stoke-on-Trent, where a 640ft-long shed could take 200 wagons under cover.

More interesting than the external appearance of these depots was the way in which they were worked: their design reflected the changes that were being brought about in logistics. At Tyneside, conveyor belts transferred goods to road vehicles. At Stoke-on-Trent, movement within the depots was by small electric trucks, and mobile cranes were employed, together with an overhead traverser crane. Palletisation of small consignments led to the use of forklift trucks on a large scale; by 1954 pallets were employed at 39 depots with a further 35 authorised or under construction. As a precursor of what was to come in the late 1960s, containers too were increasingly employed. By 1957, 36,000 were in service with a further 7,480 added that year. But they were still of small capacity, their size limited by the shortness of the traditional British railway wagon.

British Railways acquired its first two computers in 1957 to deal with payroll, but was going to outside providers for other applications, including goods. In 1958, the first work was done on the goods side when LEO, the computer of J Lyons & Co, was used to compute the minimum chargeable station-to-station distances for BR's 5,000-odd goods stations. Other new methods of speeding up the voluminous paperwork were tried, including microfilming consignment notes at Stockton-on-Tees. Increased space in the offices was given over to communications equipment, with punch-card computers and teleprinters used to dispatch consignments.

But all this was in vain. The Beeching Report of 1963 had made it clear that the future of rail freight was that the unit of freight movement must be the train rather than the individual wagon. Traditional wagonload traffic (referred to as 'sundries') was hopelessly uneconomic, and even the new marshalling yards built under the 1955 Modernisation Plan and enhanced methods of traffic

Figure 70
The Hither Green Continental Freight Depot for perishable traffic, opened in 1960, exemplifies British Railways' thinking of the period: large, unadorned sheds, clad in asbestos sheeting on a steel frame. The scale of the work is evident, with the structure over 1000ft in length. Closure came in 1987 and the site is today occupied by housing.
[John Minnis Collection]

control brought about by developments in information technology did not alter this fact. Bulk freight was the future: trains carrying one commodity from one terminal to another, be it coal, fuel oil or containers. Goods traffic of the type found today only emerged in 1966 with the launch of the Freightliner service, with long bogie wagons carrying the new large shipping containers made to international UIC standards from terminal to terminal where they were loaded directly onto trucks by cranes, with none of the manual handling of the loads at rail terminals that was previously involved. Much of the old plant – marshalling yards, goods depots, warehouses – became redundant almost immediately. All that was needed was a large open yard, with roadways running alongside the tracks and giant overhead cranes. Under the Transport Act 1968, the remaining sundries traffic was hived off to a new state-owned entity, National Carriers Ltd, which continued to use some of the old depots for a few years before increasingly turning to road transport instead of rail in the 1970s.

Traditional wagonload traffic was in a state of constant decline and finally ceased altogether in 1984. One last attempt was made to retain some of it with the introduction in 1972 of fast overnight services with new and much larger air-braked wagons. These services, branded in 1977 as Speedlink, carried cargoes such as food and drink, timber, paper, brick, steel and certain trainload and European traffic, but lost money and closed in 1991.

With these changes, the goods shed and warehouse became redundant. Most of the small country ones had already been closed by the late 1960s, and by the mid-1980s all the large ones were gone too. Those such as Peterborough, Tyneside and Grimesthorpe had very short lives (Grimesthorpe finally closed following a fire in 1985). The large, immensely valuable sites occupied by city centre depots proved irresistible to developers, and almost all were redeveloped within a few years of closure. Just two of the post-war depots, Bury St Edmunds (*see* Fig 69) – now a skating rink and Indian restaurant – and Stoke-on-Trent, survive, along with only a handful of inter-war examples.

But just as the story seemed to be coming to a close, a new generation of railway warehouses started to be constructed on greenfield sites. The Daventry International Rail Freight Terminal, whose operation commenced in the late 1990s, is an intermodal development located at junction 18 of the M1 motorway. Initial capacity was just under 0.5 million sq ft of rail-connected warehousing; subsequent expansion has seen 1.9 million sq ft of warehouse space added, with a further 7.5 million sq ft planned. The warehousing takes the form of 'big sheds' of the type usually found at motorway interchanges, and the tracks do not enter the warehouses. Similar facilities are planned elsewhere. Work began in 2015 on iPort Doncaster, located adjacent to the M18 motorway, with a 35-acre intermodal rail freight terminal and 6 million sq ft of warehousing. Again, container trains will enter the site and giant overhead cranes will remove the containers, which will then be loaded onto trucks or taken for warehousing.

So there will be a future for the successor to the traditional railway goods shed and warehouse, albeit in scarcely recognisable form, as part of an intermodal logistics system.

7 Conservation

Simon Hickman

Introduction

A conversation with an English town planner can be more baffling than trying to use schoolboy French to order a meal in a Parisian restaurant. The planner's language is sprinkled with unfamiliar words and phrases, from 'curtilage' to 'residential amenity' via a 'section 106 agreement'. In the last few years, the planner has gained another new term to add to his vocabulary – 'optimum viable use'.

Helpfully for those who struggle as much with planning-speak as they do with French waiters, the Government has published their equivalent of a foreign-language phrase book, known as Planning Practice Guidance. These online documents are designed to give guidance about how national planning policy should be applied, and definitions of those terms familiar to the planner but less so to everyone else. In the context of a heritage asset ('heritage asset' is translated as a building, monument, site, place, area or landscape identified as having a degree of significance meriting consideration in planning decisions, because of its heritage interest), the Planning Practice Guidance explains optimum viable use, stating:

> The optimum viable use may not necessarily be the most profitable one. It might be the original use, but that may no longer be economically viable or even the most compatible with the long-term conservation of the asset.

A conservation professional may well hold the view that the best use for a historic building is the one for which it was originally designed. It is often possible to sustain the original use of a historic building; a city church may close because of its declining congregation, for instance, but another faith group may be able to make use of it for the same purposes.

However, there are a number of heritage assets whose original use either no longer exists or requires a very different type of structure. The railway goods shed needs to be added to the list of such buildings as gunpowder works, whaling stations and colliery bathhouses.

The King's Cross granary, a fine example of reuse as the focal point of the University of the Arts campus. [DP149043]

Conservation

Given the relatively commonplace nature of the railway goods shed, it comes as a surprise to find that there is not one example left in the country which is in use for its original purpose as a rail transhipment point for commercial goods. Wagonload rail traffic finally died out in the 1990s and the freight (no longer 'goods'!) trains we see on today's network is either container traffic or bulk commodities, to be loaded and unloaded at modern, purpose-built terminals and not in Victorian goods sheds.

The details of extant goods sheds and warehouses given in the Gazetteer indicate an uneven pattern of survival. As one might expect, the largest railway companies tend to have the greatest number of extant structures: hence the GWR with 101, the NER with 81 and the MR with 66 have the most survivors. However, some large companies are less well represented: the GCR's total of 16 is particularly low for its size. The rate of survival is very much bound up with the type of structures put up by each company, the fate of the lines built by it and the territory it served.

The LNWR, with a total of 58 surviving sheds, has proportionately fewer survivors than its contemporaries of similar size, but it put up fewer goods sheds than some of them and was also prone to erecting wooden sheds, which have been far more susceptible to demolition. The GCR was to experience the closure of much of its main line, while both the GCR and the LNWR had many sheds in urban areas which are subject to substantial development pressures. The railway companies in the south-east lost many of their sheds to car parking and residential development. Those companies operating in agricultural areas where land values are low and where there is scope for such uses as agricultural storage or scrapyards have far more of their buildings surviving.

Attrition rates are particularly high at both ends of the scale in terms of size: small timber lock-up sheds have fared badly because they are prone to decay and the opportunity for re-use is limited, while the large sheds and multistorey warehouses in urban locations are on highly valuable sites that offer scope for comprehensive redevelopment. It has also proved difficult to attract developers willing to risk the cost of regenerating such large structures which were often in very poor condition following decades of neglect.

The remarkable, arc-shaped LNWR/MS&LR Clegg Street warehouse of 1876 at Oldham (Fig 71) survived for over 40 years after its closure in the late

1960s before being demolished in 2012. Despite being listed Grade II, its structural state had been allowed to deteriorate to the point that it was unsafe, following the collapse of various schemes for repair and reuse. With only a handful of exceptions, all the large multistorey warehouses that survive are protected by listing, but this clearly does not guarantee their survival, although the great majority are now in good repair.

Goods sheds often have historic or architectural interest. Some are listed in their own right, others as part of a station complex which is designed in a similar architectural style. They are part of a familiar and valued local scene, and if they become redundant then new uses must be found. The Planning Practice Guidance puts it succinctly: 'Putting heritage assets to a viable use is likely to lead to the investment in their maintenance necessary for their long-term conservation.'

Figure 71
A conservation failure, the Clegg Street, Oldham warehouse, in a derelict state in March 2001. [Gordon Biddle]

Reuse

The reuse of goods sheds is something that dates back to the very early days of railways. The earliest identified goods shed of all, that of 1827 on the Stockton & Darlington Railway at Darlington, was converted in 1833 to a passenger station, comprising a booking office, a waiting room and a cottage; the large goods shed at Alne (*see* Fig 7) was converted to six cottages in 1842; and the Leeds & Selby Railway goods shed at Micklefield (*see* Fig 6) was converted to a house in 1886. Conversions were not just residential: when they became redundant many goods sheds were let to outside organisations for storage. But the practice obviously gathered pace in the early 1960s as small goods yards closed and branch lines were ripped up.

In establishing the scope for new uses for historic goods sheds, the conservation professional needs to consider what is significant about the building, to identify what change might be possible. Historic England has come up with a straightforward way of evaluating significance through the four 'heritage values' set out in its 2008 publication *Conservation Principles*. To sum these up briefly, evidential value is essentially archaeological value; historical value is the ways in which past people, events and aspects of life can be connected through a place to the present; aesthetic value is the ways in which people draw sensory and intellectual stimulation from a place; and communal value is the meanings of a place for the people who relate to it, or for whom it figures in their collective experience or memory.

So how do you apply the heritage values to a goods shed to establish its significance? Let's use the 1885-built Grade II listed goods shed at Swanage, Dorset, as an example. It may have little archaeological value, but it does perhaps have a small amount of communal value, being part of a suite of station buildings associated with the well-known holiday town. There is certainly historic value as representative of a means of transporting cargo that no longer exists. But easiest to define is the building's aesthetic value. It is built in the same coursed local limestone as the nearby station and surviving engine shed. It has traditional timber doors sheltered by individual canopies with sawtoothed valances on the road side. It retains a voluminous interior, from which the simple robust timber roof structure can be seen. We can draw the heritage value of the goods shed together in a narrative, which allows us to define the significance of the building.

The Swanage goods shed has found a happy retirement as a maintenance facility for the steam engines of the Swanage Railway, which operates as a heritage line. With pick-up goods trains unlikely to make a return, it is difficult to think of a better use for the shed at Swanage. Externally it still looks as it always did, and internally its volume is retained as one single space. The relative lack of windows is an advantage to the railway for security purposes and noise reduction, while the large doors continue to provide easy means of loading.

At the other end of the Swanage branch line is Wareham station. Unlike Swanage it is part of the national network and receives a good train service as a stop on the South Western main line from Waterloo to Weymouth. Its goods shed also remains, protected by its Grade II listing, but was until recently in very poor repair and on the verge of collapse. Rescued from dereliction by Morgan Carey Architects for its own use, it has recently been superbly restored and converted to an open-plan office (Fig 72). The significance of the building has been 'sustained and enhanced', as our town planner might say.

Figure 72
The 1847 LSWR shed at Wareham, severely damaged by fire and now restored as architects' offices.
[Nigel Rigden]

Part of the success of the conversion is that open-plan interior. Allowing views to remain open through the building and up to its roof structure enables the viewer to appreciate the original design intention of the building as a single space.

The problem with reuse is that it tends to lead to loss of character. While many goods sheds, such as that at Ludlow, now turned into a brewery (Fig 73), retain much of their original external appearance when reused, very few indeed have kept the basic internal fittings of platform and jib crane. Partitioning invariably changes the appearance of the interior completely, while residential conversion often entails the insertion of extra windows, as in the conversion of the large goods shed at Ware (Fig 74). Remarkably few goods sheds have survived unaltered, even those on heritage railways where they have generally been used as workshops, locomotive sheds or railway museums. Those that have been used to inform visitors how goods traffic was handled are few indeed; among the few examples are the shed at Shildon, now part of Locomotion, which serves as an interpretive display (although the displays tend to obscure much of the shed); and the Bluebell Railway's recreation of a country goods yard at Kingscote, using the lock-up shed from Horsted Keynes. At Goathland station on the North Yorkshire Moors Railway, the goods shed has been sensitively transformed into a tea room, complete with restored open wagons used as seating areas. Perhaps the best example of interpreting a goods shed for a present-day audience is at Beamish, an open-air museum where the shed formerly at Alnwick has been reconstructed and the interior, complete with iron jib crane, is filled with the type of goods that might have been transported in its heyday.

But retention of historic character internally is not always possible in conversions, which often require multiple spaces. The goods shed at Okehampton – also a LSWR example – has been imaginatively converted to a youth hostel. Externally the building has been preserved and the contribution it makes to the environs of Okehampton station is undiminished, but internally little sense of a historic goods shed remains. However, the building is not listed – and what better place for a youth hostel than by a railway station? National planning policy encourages us to weigh any harm caused by alterations against any wider benefits. There may be some harm to the significance of the building in its subdivision, but that has to be weighed against the benefits of bringing the structure into a viable use. Whether a compartmentalised interior is the 'optimum viable use' in terms of preserving significance is a subject for debate.

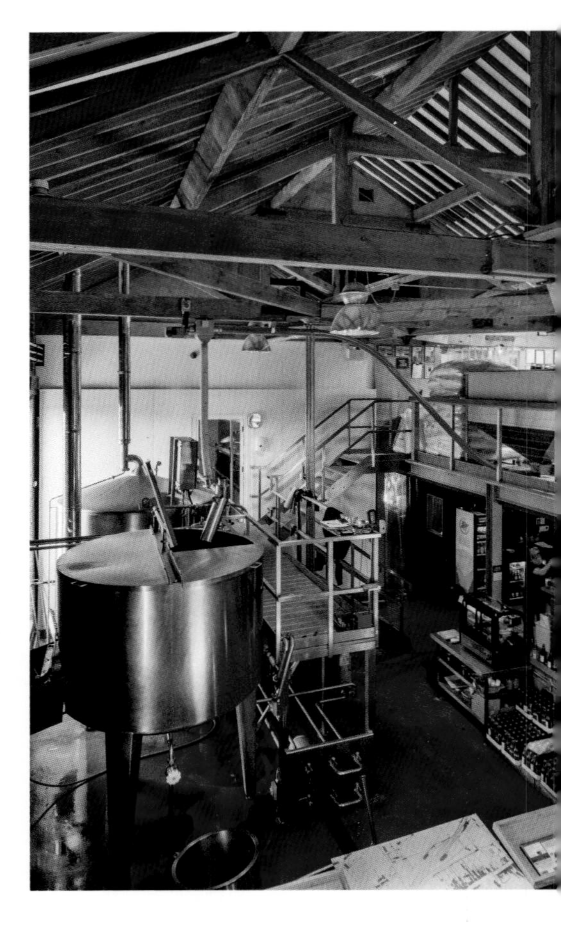

Figure 73 The interior of Ludlow's GWR/LNWR joint goods shed, now converted to a brewery. [DP173396]

Of course, keeping the interior of a goods shed unobstructed by partitions does not necessarily mean the space will automatically show its origins as industrial architecture. At Barnstaple, Victoria Road, the station has long been demolished but the large (and unlisted) goods shed remains, and has been converted to a church. Externally, a spirelet has been added to denote the new use but it remains highly legible as a railway building. Internally, however, the church's style of worship required a blank box, and there are now few signs of the shed's origins as a railway building.

At the other end of the spectrum is a giant amongst goods depots, the former GNR yards at King's Cross, London. Perhaps a victim of benign neglect, following cessation of railway use in the early 1970s the huge complex of buildings was used for light industrial purposes for many years, until the rapid resurgence of the King's Cross area was kickstarted by the conversion of St Pancras Station into the London terminal of the Channel Tunnel Rail Link. In a more prosperous area, the buildings of the King's Cross depot might have been lost long before anyone thought about listing them.

Figure 74
Ware was provided with a large goods shed at its opening in 1843 to deal with malt and other agricultural traffic. Reuse as loft-type apartments has resulted in what were formerly lunettes above recessed brick panels being replaced by large openings extending to plinth level. Further small rectangular windows have been introduced just below the eaves to light an extra inserted floor.
[DP173933]

The King's Cross area is railway architecture on an epic scale. Lewis Cubitt's 1852 station is complemented by his goods complex on land to the north, where a heroic granary building, now Grade II listed (*see* Chapter 7 opener) fronts a complex of structures designed to handle every type of goods traffic imaginable. As discussions about the future of the area began in the 1990s, London's famous arts institution, St Martin's College (now the University of the Arts), stepped forward to suggest it might be able to use the Eastern Goods Yards for a new university campus.

The University of the Arts, its architects Stanton Williams and the site developer Argent developed into a formidable partnership, in conjunction with Camden Council and English Heritage. A conversion of the complex skilfully combined radical intervention with scholarly conservation, and the result is truly marvellous. To the rear of the granary, a shed where wagons were marshalled for onward travel was flanked by two long, linear, brick transit sheds where goods were unloaded. The heavily altered and woebegone assembly shed was removed to create space for a huge concrete structure housing lecture theatres and teaching facilities, the scale of which is only externally apparent at the north end where a lecture theatre cantilevers over a new public space between the twin gables of the transit sheds.

The creative environment of the arts college allowed for historic fabric to be conserved sensitively, with the scars and blackening of the building's industrial past preserved in a very honest manner. The vibrancy of the student population and busy pedestrian movement across the site animates the development. The result is one of the most spectacular contrasts between old and new that can be seen in England.

Was the significance of the buildings preserved, however? Despite the loss of the assembly shed, the linearity of the complex remains. Its history is celebrated in the 'as found' approach taken to its conservation. The aesthetic of the building has, in the minds of many, been enhanced by the interventions.

It is interesting to consider that 20 years ago no developer would have considered investing in King's Cross, yet today one of the world's most well-known companies, Google, is currently constructing its new headquarters there. Predicting future changes in society is difficult, but as the King's Cross story demonstrates, there is nearly always a future for a large adaptable building located close to a transport hub.

Passenger traffic on our rail network is booming, and the role of stations is changing. The relationship between the employer and the employee has matured and the latter is often free to work in a way that suits him or her.

A modern employee may no longer have an office, but still needs to use one occasionally to meet people. The goods shed offers a shell ready to be converted to offices that can be rented by the hour, or to a depot to pick up something bought from the Internet, at a time and place convenient to the commuter. As more people turn to the bicycle as a means of getting to the station, the goods shed offers potential for excellent cycle storage facilities, often right next to the station platforms.

Two of the basic principles of conservation advice are to retain as much historic fabric as possible and to allow for reversibility in alterations. The most successful examples of goods shed conversions are where those philosophies have been applied, allowing yesterday's goods shed to be today's industrial unit but tomorrow's college. To preserve our past – and to keep a heritage asset in its optimum viable use – we must be flexible about its future.

Gazetteer

The Gazetteer is an attempt to note every extant goods shed or warehouse in England at the time of publication. It is possible that one or two buildings may have been demolished since going to press. Considerable trouble has been taken to ensure that the Gazetteer is as complete as possible but it is almost inevitable that an occasional survivor may have slipped through the net, and the author would welcome details of these. In particular, there may be more examples of goods lock-ups which, as they are often mounted on platforms of passenger stations, are not always easy to identify.

Entries in the company column refer to the owning company at the time of the 1923 railway grouping, not necessarily the company that actually built the structure. Dates are given where known, but research into railway buildings on a company basis has been patchy, to say the least, with some companies, such as the NER and the M&GNR, having been the subject of considerable study and others, such as the LNWR, very little. A fuller version of the database on which this Gazetteer is based is available on the Historic England website, www.HistoricEngland.org.uk.

Company	Location	Listed	Date	County	NGR
Birkenhead J	Bromborough			Cheshire	SJ3438480841
Birkenhead J	Ellesmere Port			Cheshire	SJ4044176538
Birkenhead J	Frodsham		1850	Cheshire	SJ5188377882
Cambrian	Oswestry			Shropshire	SJ2934429741
Cambrian	Weston Wharf, Oswestry			Shropshire	SJ2984327606
CLC	Birkenhead warehouse		1889	Merseyside	SJ3268089275
CLC	Cheadle		1865	Greater Manchester	SJ8549189408
CLC	Mickle Trafford			Cheshire	SJ4461669171
CLC	Warrington warehouse	II	1897	Cheshire	SJ6082588610
CV&HR	Birdbrook			Essex	TL7161441942
FR	Arnside			Cumbria	SD4619578802
FR	Barrow-in-Furness			Cumbria	SD1931269146
FR	Bootle			Cumbria	SD0937489364
FR	Cark & Cartmel			Cumbria	SD3656776214
FR	Drigg			Cumbria	SD0636898868
FR	Grange-over-Sands			Cumbria	SD4127078189
FR	Haverthwaite		1869	Cumbria	SD3494584219
FR	Millom			Cumbria	SD1724780171
FR	Ravenglass			Cumbria	SD0854596494
FR	Seascale			Cumbria	NY0377201042
FR	Silecroft			Cumbria	SD1315281934
FR	Torver			Cumbria	SD2837094116
FR	Ulverston			Cumbria	SD2853277929
FR/LNWR	Beckermet			Cumbria	NY0157406598
FR/LNWR	Woodend			Cumbria	NY0092013029
FR/MR	Borwick		1867	Lancashire	SD5364672880
FR/MR	Melling		1867	Lancashire	SD6011471514

Company	Location	Listed	Date	County	NGR
GCR	Brackley		1899	Northamptonshire	SP5905237858
GCR	Glossop			Derbyshire	SK0346494170
GCR	Grimsby			Lincolnshire	TA2738109956
GCR	Hadfield			Derbyshire	SK0241696057
GCR	Kirton Lindsey			Lincolnshire	SK9335799563
GCR	Langworth			Lincolnshire	TF0526975990
GCR	Lincoln warehouse		1907	Lincolnshire	SK9721270946
GCR	Loughborough			Leicestershire	SK5431419421
GCR	Manchester Ducie Street warehouse	II	1867	Greater Manchester	SJ8476198073
GCR	Mottram			Greater Manchester	SJ9914893800
GCR	Northorpe			Lincolnshire	SK9052596583
GCR	Penistone			Yorkshire	SE2443803398
GCR	Quorn & Woodhouse		1899	Leicestershire	SK5498016069
GCR	Rothley		1899	Leicestershire	SK5693212091
GCR	Swithland Sidings		1899	Leicestershire	SK5643113169
GCR	Wigan			Lancashire	SD5883705258
GCR/MR	Reddish North			Greater Manchester	SJ8977994711
GER	Acle		1881	Norfolk	TG3974109938
GER	Attleborough		1845	Norfolk	TM0520095034
GER	Bealings		1859	Suffolk	TM2304547399
GER	Black Bank			Cambridgeshire	TL5322985384
GER	Braughing			Hertfordshire	TL3901324238
GER	Burnham Market granary		1866	Norfolk	TF8342541980
GER	Bury St Edmunds		1954	Suffolk	TL8524665088
GER	Chappel & Wakes Colne		1891	Essex	TL8981328888
GER	Clare	II	1865	Suffolk	TL7707045154
GER	Dereham			Norfolk	TF9937613100

Company	Location	Listed	Date	County	NGR
GER	Dunham		1848	Norfolk	TF8681813256
GER	Felsted lock-up			Essex	TL6644321224
GER	Frinton lock-up			Essex	TM2351220427
GER	Geldeston		1863	Norfolk	TM3866591688
GER	Glemsford			Suffolk	TL8314046517
GER	Harwich Town		1865	Essex	TM2595332527
GER	Hatfield Peverel lock-up			Essex	TL7891112207
GER	Higham			Suffolk	TL7471666086
GER	Histon			Cambridgeshire	TL440906262
GER	Hockley lock-up			Essex	TQ8426392752
GER	Holme Hale		1875	Norfolk	TF8776607002
GER	Homersfield			Norfolk	TM2822985975
GER	Leiston		1859	Suffolk	TM4423762893
GER	Lingwood granary		1884	Norfolk	TG3610208370
GER	Little Walsingham			Norfolk	TF9318936810
GER	Maldon West		1889	Essex	TL8419606112
GER	March			Cambridgeshire	TL4212797734
GER	Marks Tey		1865	Essex	TL9166323962
GER	Mildenhall		1885	Cambridgeshire	TL7074273991
GER	Narborough		1846	Norfolk	TF7424613297
GER	North Walsham			Norfolk	TG2817929802
GER	Oakington			Cambridgeshire	TL4180965229
GER	Peterborough East	II	1847	Cambridgeshire	TL1952698005
GER	Reepham granary			Norfolk	TG1014023488
GER	Rochford lock-up			Essex	TQ8731390458
GER	Rochford		1889	Essex	TQ8739890428
GER	Roydon		1848	Essex	TL4055610503

Company	Location	Listed	Date	County	NGR
GER	Saxmundham			Suffolk	TM3852263210
GER	Smeeth Road			Norfolk	TF5203209412
GER	Snettisham			Norfolk	TF6767933444
GER	Stoke Ferry		1882	Norfolk	TL7066699604
GER	Swaffham		1847	Norfolk	TF8190809495
GER	Takeley lock-up			Essex	TL5620221070
GER	Thetford		1912	Norfolk	TL8672383658
GER	Three Horse Shoes			Cambridgeshire	TL3359296887
GER	Tivetshall			Norfolk	TM1581188004
GER	Ware	II		Hertfordshire	TL3607013983
GER	Wivenhoe	II	1903	Essex	TM0361721676
GER	Woodbridge			Suffolk	TM2731448751
GER	Wymondham	II	1845	Norfolk	TG1144000997
GER	Yaxham			Norfolk	TG0030810147
GNR	Aby & Claythorpe			Lincolnshire	TF4123379118
GNR	Algarkirk & Sutterton		1848	Lincolnshire	TF2907634361
GNR	Ancaster		1857	Lincolnshire	SK9834144385
GNR	Ashwell & Morden			Cambridgeshire	TL2988938681
GNR	Aswarby & Scredington			Lincolnshire	TF0887840976
GNR	Bardney	II	1848	Lincolnshire	TF1133569080
GNR	Bingham		1850	Nottinghamshire	SK7018540186
GNR	Boston		1903	Lincolnshire	TF3243243272
GNR	Burgh-le-Marsh			Lincolnshire	TF4766866543
GNR	Caythorpe			Lincolnshire	SK9475048369
GNR	Derby Friargate warehouse	II	1878	Derbyshire	SK3457136198
GNR	Eastville			Lincolnshire	TF4058056870
GNR	Firsby			Lincolnshire	TF4594464221

Company	Location	Listed	Date	County	NGR
GNR	Foxton		1883	Cambridgeshire	TL4082348729
GNR	Halifax warehouse	II	1885	West Yorkshire	SE0966524746
GNR	Halton Holgate			Lincolnshire	TF4158664487
GNR	Harmston			Lincolnshire	SK9630462153
GNR	Heckington		1859	Lincolnshire	TF1471043570
GNR	Horncastle		1855	Lincolnshire	TF2540569474
GNR	Ilkeston			Derbyshire	SK4642342856
GNR	Keighley	II	*c* 1884	West Yorkshire	SE0642641171
GNR	Keighley warehouse			West Yorkshire	SE0659441086
GNR	King's Cross granary	II	1850	Greater London	TQ3015183606
GNR	King's Cross Midland	II	*c* 1850	Greater London	TQ3024483586
GNR	Leadenham		1867	Lincolnshire	SK9572252910
GNR	Letchworth			Hertfordshire	TL2229933103
GNR	Littleworth			Lincolnshire	TF2080415158
GNR	Low Moor	II	1893	West Yorkshire	SE1674528378
GNR	Manchester Deansgate warehouse	II*	1898	Greater Manchester	SJ8357097872
GNR	Morley Top			West Yorkshire	SE2626527405
GNR	Navenby		1867	Lincolnshire	SK9759857973
GNR	Nottingham warehouses	II	1857	Nottinghamshire	SK5813639336
GNR	Retford			Nottinghamshire	SK6980680836
GNR	Rippingale		1872	Lincolnshire	TF1151528280
GNR	Shepreth		1851	Cambridgeshire	TL3919548160
GNR	Spilsby			Lincolnshire	TF4012065780
GNR	Stamford East	II	1856	Lincolnshire	TF0350806926
GNR	Stanningley			West Yorkshire	SE2218734353
GNR	Tattershall	II	1848	Lincolnshire	TF2041056837
GNR	Tempsford			Bedfordshire	TL1800354031

Company	Location	Listed	Date	County	NGR
GNR	Tuxford		1852	Nottinghamshire	SK7474071312
GNR	Willoughby			Lincolnshire	TF4669371886
GNR	Wilsden		1886	West Yorkshire	SE0760835351
GNR/GCR Jt	South Elmsall			Yorkshire	SE4742211134
GNR/GER Jt	Beckingham		1867	Nottinghamshire	SK7864989608
GNR/GER Jt	Cowbit		1867	Lincolnshire	TF2655518157
GNR/GER Jt	Digby		1882	Lincolnshire	TF0875154790
GNR/GER Jt	Donington Road		1882	Lincolnshire	TF2016735104
GNR/GER Jt	Finningley		1867	South Yorkshire	SK6730999698
GNR/GER Jt	Gosberton		1882	Lincolnshire	TF2229730046
GNR/GER Jt	Haxey & Epworth		1867	Lincolnshire	SK7679697383
GNR/GER Jt	Helpringham		1882	Lincolnshire	TF1324540690
GNR/GER Jt	Misterton		1867	Nottinghamshire	SK7761994105
GNR/GER Jt	Murrow		1866	Cambridgeshire	TF3693406370
GNR/GER Jt	Nocton & Dunston		1882	Lincolnshire	TF0571263018
GNR/GER Jt	Pinchbeck		1882	Lincolnshire	TF2354525676
GNR/GER Jt	Postland		1867	Lincolnshire	TF2921812407
GNR/GER Jt	Potter Hanworth		1882	Lincolnshire	TF0453166332
GNR/GER Jt	Scopwick & Timberland		1882	Lincolnshire	TF0940158382
GNR/GER Jt	Stow Park			Lincolnshire	SK8569881414
GNR/GER Jt	Twenty Foot River			Cambridgeshire	TF4061800908
GNR/GER Jt	Walkeringham		1867	Nottinghamshire	SK7770092613
GNR/LNWR Jt	Harby & Stathern		1879	Leicestershire	SK7602630645
GNR/LNWR Jt	Tilton		1879	Leicestershire	SK7604605828
GWR	Abbotsbury		1885	Dorset	SY5830085301
GWR	Acton			Greater London	TQ2023081239
GWR	Ashburton	II	1872	Devon	SX7568669705

Company	Location	Listed	Date	County	NGR
GWR	Avonwick lock-up		1893	Devon	SX7178657549
GWR	Axbridge		1869	Somerset	ST4322454681
GWR	Barnstaple Victoria Road			Devon	SS5666232731
GWR	Bath		1877	Somerset	ST7453864400
GWR	Bewdley		1862	Worcestershire	SO7930175289
GWR	Bishops Lydeard		1862	Somerset	ST1639328979
GWR	Bishops Nympton & Molland		1873	Devon	SS7880026476
GWR	Bordesley warehouse		1931	West Midlands	SP0866085711
GWR	Bourne End		1873	Buckinghamshire	SU8943987224
GWR	Bovey Tracey		1866	Devon	SX8111678329
GWR	Bridgnorth		1862	Shropshire	SO7150692680
GWR	Bristol, Canons Marsh	II	1906	Avon	ST5837472552
GWR	Broadway		1904	Worcestershire	SP0871238104
GWR	Buckfastleigh		1872	Devon	SX7459866308
GWR	Burghclere		1885	Hampshire	SU4698157814
GWR	Cadeleigh		1885	Devon	SS9381207581
GWR	Camborne			Cornwall	SW6471439618
GWR	Castle Cary			Somerset	ST6334433454
GWR	Cheddar		1869	Somerset	ST4536053269
GWR	Coleford		1883	Gloucestershire	SO5765510490
GWR	Corsham			Wiltshire	ST8702269695
GWR	Culkerton		1889	Gloucestershire	ST9242896144
GWR	Didcot transfer shed	II	1863	Oxfordshire	SU5231691261
GWR	Dulverton		1873	Somerset	SS9268025501
GWR	Dunster	II	1874	Somerset	SS9961844732
GWR	Dymock			Gloucestershire	SO6992030989
GWR	East Anstey		1873	Devon	SS8675126132

Company	Location	Listed	Date	County	NGR
GWR	Edington & Bratton		1900	Wiltshire	ST9232253761
GWR	Exeter St David's transfer shed	II	1860	Devon	SX9112093676
GWR	Fencote			Herefordshire	SO6008258945
GWR	Flax Bourton lock-up			Somerset	ST5130369762
GWR	Gotherington lock-up		1906	Gloucestershire	SO9747729866
GWR	Hatch		1866	Somerset	ST3053320372
GWR	Hele & Bradninch	II		Devon	SS9951002288
GWR	Helston			Cornwall	SW6628228056
GWR	Hereford		1855	Herefordshire	SO5134340701
GWR	High Wycombe (converted from passenger station)		1854	Buckinghamshire	SU8684093053
GWR	Hodnet		1867	Shropshire	SJ6213927878
GWR	Horrabridge			Devon	SX5111669401
GWR	Hullavington		1903	Wiltshire	ST9008282807
GWR	Ilminster		1866	Somerset	ST3483314872
GWR	Ivybridge			Devon	SX6318256613
GWR	Keynsham		1911	Somerset	ST6598168688
GWR	Kidderminster			Worcestershire	SO8380176141
GWR	Kidderminster			Worcestershire	SO8374876265
GWR	Kingsbridge		1893	Devon	SX7307444129
GWR	Kington		1857	Herefordshire	SO3040157014
GWR	Leamington Spa			Warwickshire	SP3144465251
GWR	Liverpool Pier Head		1890	Merseyside	SJ3399690023
GWR	Loddiswell lock-up		1893	Devon	SX7307848373
GWR	Lodge Hill			Somerset	ST4986148440
GWR	Lye			Worcestershire	SO9235084669
GWR	Manchester Lower Byrom Street warehouse	II	1880	Greater Manchester	SJ8317297868

Company	Location	Listed	Date	County	NGR
GWR	Melksham warehouse			Wiltshire	ST8999264619
GWR	Minehead	II	1874	Somerset	SS9749046325
GWR	Morebath		1873	Devon	SS9760824416
GWR	Moretonhampstead		1866	Devon	SX7570385706
GWR	Much Wenlock			Shropshire	SJ6215000094
GWR	Newland			Gloucestershire	SO5546410471
GWR	Newton Abbot		1911	Devon	SX8658371858
GWR	Paignton			Devon	SX8894060585
GWR	Pembridge		1857	Herefordshire	SO3882259077
GWR	Perranwell		1863	Cornwall	SW7806439786
GWR	Portesham		1885	Dorset	SY6048385515
GWR	Rednal & West Felton			Shropshire	SJ3530227537
GWR	Ross-on-Wye		1855	Herefordshire	SO6046624370
GWR	Rowden Mill lock-up		1897	Herefordshire	SO6269656572
GWR	Sandford & Banwell	II	1869	Somerset	ST4162759507
GWR	Shrewsbury			Shropshire	SJ4937313180
GWR	Slough			Berkshire	SU9745380295
GWR	South Brent		1893	Devon	SX6977960276
GWR	South Molton		1873	Devon	SS7183627062
GWR	Southall			Greater London	TQ1289379886
GWR	Starcross lock-up		1893	Devon	SX9769281902
GWR	Staverton			Devon	SX7838763780
GWR	Stoke Edith			Herefordshire	SO6141841327
GWR	Stroud	II	1845	Gloucestershire	SO8508005030
GWR	Sutton Scotney			Hampshire	SU4653339549
GWR	Swindon			Wiltshire	SU1598485645
GWR	Tetbury		1889	Gloucestershire	ST8932893234

Company	Location	Listed	Date	County	NGR
GWR	Tettenhall		1913	Worcestershire	SO8909899883
GWR	Tiverton			Devon	SS9613012665
GWR	Toddington		1904	Gloucestershire	SP0498932131
GWR	Torre			Devon	SX9025664979
GWR	Upwey		1886	Dorset	SY6667883632
GWR	Venn Cross		1873	Somerset	ST0323724614
GWR	Watchet		1862	Somerset	ST0714543225
GWR	Wellington			Somerset	ST1308121290
GWR	Wells Tucker Street	II	1870	Somerset	ST5432845573
GWR	Williton	II	1862	Somerset	ST0853741587
GWR	Wilton		1856	Wiltshire	SU0989832031
GWR	Winchcombe		1905	Gloucestershire	SP0275729699
GWR	Winchester Chesil			Hampshire	SU4865128742
GWR	Winterbourne		1903	Gloucestershire	ST6529379939
GWR	Witney		1862	Oxfordshire	SP3581009024
GWR	Wiveliscombe		1871	Devon	ST0849227635
GWR	Wookey			Somerset	ST5312246338
GWR	Worcester			Worcestershire	SO8589455170
GWR	Yeovil, Clifton Maybank transfer shed	II	1864	Somerset	ST5699414018
GWR/LNWR	Craven Arms			Shropshire	SO4314883210
GWR/LNWR	Ludlow		1852	Shropshire	SO5124675184
GWR/LNWR	Pontesbury		1861	Shropshire	SJ3975906331
GWR/LNWR	Woofferton		1853	Shropshire	SO5142668326
H&BR	Carlton		1885	Yorkshire	SE6469325055
H&BR	Drax		1885	Yorkshire	SE6675426414
H&BR	Hull Neptune Street			East Yorkshire	TA0866527763
L&BR	Lynton		1898	Devon	SS7191248775

Company	Location	Listed	Date	County	NGR
L&CR	St Cleer			Cornwall	SX2508368439
L&YR	Atherton			Lancashire	SD6824903671
L&YR	Baxenden		1848	Lancashire	SD7777325674
L&YR	Blackburn Bolton Road			Lancashire	SD6802827299
L&YR	Blackrod			Lancashire	SD6238710685
L&YR	Brooksbottom			Greater Manchester	SD7931214980
L&YR	Bury	II	1846	Greater Manchester	SD8019510931
L&YR	Chatburn			Lancashire	SD7639243595
L&YR	Cherry Tree		1846	Lancashire	SD6587526471
L&YR	Clayton West		1879	Yorkshire	SE2577211232
L&YR	Denby Dale		1850	Yorkshire	SE2241708557
L&YR	Ewood Bridge			Lancashire	SD7971820874
L&YR	Halifax warehouse	II	1849	West Yorkshire	SE0958624592
L&YR	Heap Bridge			Greater Manchester	SD8283810545
L&YR	Heywood	II	1841–3	Greater Manchester	SD8629410304
L&YR	Holmfirth			West Yorkshire	SE1447508610
L&YR	Honley		1850	Yorkshire	SE1454412499
L&YR	Horbury & Ossett		1902	Yorkshire	SE2850318156
L&YR	Kirkham & Wesham			Lancashire	SD4158632663
L&YR	Liversedge		1848	West Yorkshire	SE2024523779
L&YR	Midge Hall			Lancashire	SD5096723190
L&YR	Mill Hill			Lancashire	SD6707226718
L&YR	New Hey cotton shed		1913	Greater Manchester	SD9386211534
L&YR	Ravensthorpe			Yorkshire	SE2279219998
L&YR	Rawtenstall			Lancashire	SD8090622475
L&YR	Rochdale			Lancashire	SD9026912929
L&YR	Shepley			Yorkshire	SE1970110276

Company	Location	Listed	Date	County	NGR
L&YR	Simonstone		1885	Lancashire	SD7753233645
L&YR	Summerseat	II	1846	Greater Manchester	SD7943814600
L&YR	Waterfoot warehouse			Lancashire	SD8306321804
LB&SCR	Angmering		1853	Sussex	TQ0657002988
LB&SCR	Arundel	II	1863	Sussex	TQ0236706273
LB&SCR	Baynards		1865	Surrey	TQ0764735100
LB&SCR	Billingshurst		1859	Sussex	TQ0881925123
LB&SCR	Bognor Regis		1902	Sussex	SZ9351499452
LB&SCR	Burgess Hill		1889	Sussex	TQ3156518697
LB&SCR	Chichester		1881	Sussex	SU8579604346
LB&SCR	Christ's Hospital		1902	Sussex	TQ1476429142
LB&SCR	Cooksbridge		1854	Sussex	TQ4005713500
LB&SCR	Crowborough		1906	Sussex	TQ5313329372
LB&SCR	Earlswood		1855	Surrey	TQ2785449671
LB&SCR	Eastbourne			Sussex	TV6091399191
LB&SCR	Edenbridge Town	II	1888	Kent	TQ4458946485
LB&SCR	Fittleworth lock-up		1891	Sussex	TQ0072318119
LB&SCR	Hartfield		1866	Sussex	TQ4802636200
LB&SCR	Hassocks		1853	Sussex	TQ3035615468
LB&SCR	Hayling Island		1900	Hampshire	SZ7096199767
LB&SCR	Horley	II	1841	Surrey	TQ2863942979
LB&SCR	Horsted Keynes lock-up		1890	Sussex	TQ3708529103
LB&SCR	Isfield lock-up		1898	Sussex	TQ4522617178
LB&SCR	Norwood Junction		1865	Greater London	TQ3399468190
LB&SCR	Pulborough		1859	Sussex	TQ0429918615
LB&SCR	Singleton	II	1881	Sussex	SU8685213137
LB&SCR	Ewell East		1851	Surrey	TQ2253862139

Company	Location	Listed	Date	County	NGR
LCDR	Adisham		1861	Kent	TR2341853946
LCDR	Bearsted	II	1884	Kent	TQ7980656134
LCDR	Canterbury East			Kent	TR1473457262
LCDR	Clapham			Greater London	TQ2976575741
LCDR	Faversham	II		Kent	TR0216360972
LCDR	Harrietsham		1884	Kent	TQ8678352839
LCDR	Herne Bay		1861	Kent	TR1694567418
LCDR	Shoreham		1862	Kent	TQ5258961517
LCDR	Teynham			Kent	TQ9566363107
LCDR (SR)	Margate			Kent	TR3506470333
LDECR	Doddington & Harby		1897	Nottinghamshire	SK8783971451
LDECR	Ollerton		1896	Nottinghamshire	SK6499966838
LNWR	Batley			West Yorkshire	SE2495223878
LNWR	Bay Horse			Lancashire	SD4926452899
LNWR	Berkhamsted			Hertfordshire	SP9964907969
LNWR	Birkenhead, Canning Street warehouse			Merseyside	SJ3222189512
LNWR	Broome		1860	Shropshire	SO3994980982
LNWR	Bucknell			Shropshire	SO3575473745
LNWR	Calveley transhipment shed			Cheshire	SJ5922358694
LNWR	Camden Town warehouse	II	1905	London	TQ2859584092
LNWR	Castle Ashby & Earls Barton			Northamptonshire	SP8591761749
LNWR	Chillington Interchange Basin transhipment shed	II		West Midlands	SO9255898048
LNWR	Chorley			Lancashire	SD5876217881
LNWR	Cromford High Peak Wharf transhipment shed	AM		Derbyshire	SK3143655761
LNWR	Dukinfield			Greater Manchester	SJ9364398215

Company	Location	Listed	Date	County	NGR
LNWR	Earlestown			Lancashire	SJ5749394966
LNWR	Elmesthorpe			Leicestershire	SP4699595830
LNWR	Golborne			Lancashire	SJ6057998023
LNWR	Hemel Hempstead			Hertfordshire	TL0442905897
LNWR	Hincaster Junction explosives shed			Cumbria	SD5121884828
LNWR	Hopton Heath			Shropshire	SO3802077304
LNWR	Kendal			Cumbria	SD5187993228
LNWR	Kings Langley			Hertfordshire	TL0801402158
LNWR	Leicester warehouse		1898	Leicestershire	SK5962504775
LNWR	Lichfield City			Staffordshire	SK1197609257
LNWR	Longcliffe			Derbyshire	SK2257455693
LNWR	Lord's Bridge		1862	Cambridgeshire	TL3957754445
LNWR	Loughborough Derby Road		1883	Leicestershire	SK5288020124
LNWR	Manchester Grape St warehouse	II	1869	Greater Manchester	SJ8307597946
LNWR	Manchester Liverpool Road warehouse	I	1830	Greater Manchester	SJ8298297890
LNWR	Manchester Liverpool Road transit shed	II	*c* 1855	Greater Manchester	SJ8315597816
LNWR	Micklehurst			Greater Manchester	SD9775201498
LNWR	Millbrook			Greater Manchester	SD9766000176
LNWR	Milnthorpe explosives shed			Cumbria	SD5130081527
LNWR	Narborough			Leicestershire	SP5403097312
LNWR	Oakengates		1861	Shropshire	SJ6980010881
LNWR	Old North Road		1862	Cambridgeshire	TL3161054603
LNWR	Oldham			Greater Manchester	SD9350804970
LNWR	Penkridge			Staffordshire	SJ9212214780
LNWR	Ravensthorpe			West Yorkshire	SE2279219996
LNWR	Rockingham			Leicestershire	SP8655793147

Company	Location	Listed	Date	County	NGR
LNWR	Rugeley Trent Valley			Staffordshire	SK0485319156
LNWR	Sedburgh		1861	Cumbria	SD6425191954
LNWR	Shrewsbury			Shropshire	SJ4960613141
LNWR	Stafford	II	1880	Staffordshire	SJ9214522554
LNWR	Staveley			Cumbria	SD4702897999
LNWR	Steeplehouse			Derbyshire	SK2885255457
LNWR	Stockport Heaton Norris warehouse	II	1877	Greater Manchester	SJ8887290852
LNWR	Wansford			Cambridgeshire	TL0921597959
LNWR	Waverton	II	1898	Cheshire	SJ4495363582
LNWR	Wednesbury			West Midlands	SO9847794653
LNWR	Welton			Northamptonshire	SP5972968030
LNWR	Whitchurch			Shropshire	SJ5496741430
LNWR	Wolverhampton Mill Street	II	1852	Staffordshire	SO9217098740
LNWR	Wrenbury			Cheshire	SJ6017147090
LNWR/LYR	Huddersfield new warehouse	II	1885	Yorkshire	SE1425816826
LNWR/LYR	Huddersfield old warehouse	II	1869	Yorkshire	SE1424116904
LNWR/MR	Longton warehouse			Staffordshire	SJ9069943810
LNWR/MR	Market Bosworth		1873	Leicestershire	SK3927102979
LNWR/MR	Measham		1873	Leicestershire	SK3335711828
LNWR/MR	Snarestone			Leicestershire	SK3408709129
LNWR/MR	Stoke Golding		1873	Leicestershire	SP3918497374
LSWR	Alresford		1865	Hampshire	SU5880132464
LSWR	Ashbury		1879	Cornwall	SX4829896305
LSWR	Bere Alston		1890	Devon	SX4407467442
LSWR	Bere Ferrers		1890	Devon	SX4525163517
LSWR	Bow			Devon	SX7160599993
LSWR	Braunton		1874	Devon	SS4867036522

Company	Location	Listed	Date	County	NGR
LSWR	Brentford			Middlesex	TQ1753277938
LSWR	Bridestowe		1874	Devon	SX5224987244
LSWR	Broad Clyst		1860	Devon	SX9920995147
LSWR	Brockenhurst			Hampshire	SU3014602050
LSWR	Colyton		1868	Devon	SY2516294033
LSWR	Corfe Castle		1885	Dorset	SY9620982039
LSWR	Crewkerne		1860	Somerset	ST4543808515
LSWR	Eggesford			Devon	SS6820111497
LSWR	Esher			Surrey	TQ1476365827
LSWR	Farnham			Surrey	SU8431246496
LSWR	Hampton Court			Surrey	TQ1537468267
LSWR	North Tawton		1865	Devon	SS6652400034
LSWR	Okehampton			Devon	SX5922594413
LSWR	Port Isaac Road		1895	Cornwall	SX0396878811
LSWR	Romsey		1847	Hampshire	SU3568521562
LSWR	Semley		1860	Wiltshire	ST8746226784
LSWR	Sherborne		1860	Dorset	ST6391316085
LSWR	Sidmouth		1874	Devon	SY1211988630
LSWR	Southampton Terminus	II		Hampshire	SU4270511134
LSWR	St Kew Highway		1895	Cornwall	SX0306175170
LSWR	Swanage	II	1885	Dorset	SZ0279678914
LSWR	Templecombe		1860	Somerset	ST7072722498
LSWR	Tresmeer		1892	Cornwall	SX2221088529
LSWR	Wadebridge			Cornwall	SW9917372176
LSWR	Wareham	II	1847	Dorset	SY9209488219
LSWR	Whitstone & Bridgerule			Devon	SS2698101442
LSWR	Winchester			Hampshire	SU4772629965

Company	Location	Listed	Date	County	NGR
LSWR	Witley			Surrey	SU9481837918
LSWR	Wool		1877	Dorset	SY8448786905
LSWR (SR)	Exmouth			Devon	SX9992281108
M&CR	Aspatria			Cumbria	NY1443641328
M&CR	Carlisle Crown Street			Cumbria	NY4036555267
M&CR	Dalston			Cumbria	NY3662250591
M&CR	Maryport			Cumbria	NY0349935931
M&CR	Wigton			Cumbria	NY2520648805
M&GNR	East Rudham		1916	Norfolk	TF8399426396
M&GNR	Fleet		1862	Lincolnshire	TF3919724272
M&GNR	Lenwade		1882	Norfolk	TG1047818441
M&GNR	Melton Constable		1882	Norfolk	TG0432033061
M&GNR	North Drove			Lincolnshire	TF2090121239
M&GNR	South Lynn		1931	Norfolk	TF6140818387
M&GNR	Whitwell & Reepham		1882	Norfolk	TG0915421587
M&GNR	Wryde			Cambridgeshire	TF3169004908
Met Rly	Vine Street warehouse			Greater London	TQ3143882120
MR	Ambergate	II		Derbyshire	SK3503251605
MR	Ampthill			Bedfordshire	TL0228737145
MR	Appleby		1876	Cumbria	NY6884020490
MR	Armathwaite		1876	Cumbria	NY5053146485
MR	Ashwell			Rutland	SK8611113887
MR	Bakewell			Derbyshire	SK2218669153
MR	Basford			Nottinghamshire	SK5528742875
MR	Beckford		1864	Worcestershire	SO9811735587
MR	Bedford			Bedfordshire	TL0443949446
MR	Beeston			Nottinghamshire	SK5349736353

Company	Location	Listed	Date	County	NGR
MR	Bingley warehouse	II	1900	West Yorkshire	SE1066339381
MR	Bitton		1869	Gloucestershire	ST6697070382
MR	Burton upon Trent granary	II		Staffordshire	SK2421123332
MR	Burton upon Trent bonded warehouse	II	1901	Stafffordshire	SK2476224093
MR	Calverley & Rodley			West Yorkshire	SE2222437089
MR	Caton			Lancashire	SD5310764814
MR	Charfield		1844	Gloucestershire	ST7238492286
MR	Coaley	II	1856	Gloucestershire	SO7494602063
MR	Derby St Mary's	II		Derbyshire	SK3553736953
MR	Derby St Mary's granary	II		Derbyshire	SK3552836883
MR	Derby St Mary's bonded warehouse			Derbyshire	SK3579436977
MR	Dewsbury			West Yorkshire	SE2430421239
MR	Earby			Lancashire	SD9044246476
MR	Eardisley			Herefordshire	SO3116748542
MR	Edmondthorpe & Wymondham		1894	Leicestershire	SK8501319031
MR	Farnsfield		1871	Nottinghamshire	SK6433957237
MR	Guiseley			West Yorkshire	SE1878742268
MR	Halton			Lancashire	SD5036964587
MR	Haworth			West Yorkshire	SE0345637057
MR	Hay on Wye			Herefordshire	SO2309742873
MR	Helpston granary	II	1860	Cambridgeshire	TF1327905421
MR	Hucknall			Nottinghamshire	SK5395149393
MR	Ingrow		1867	Yorkshire	SE1327905421
MR	Irchester			Northamptonshire	SP9313566171
MR	Isham & Burton Latimer			Northamptonshire	SP8875074745
MR	Kegworth			Nottinghamshire	SK5006326753
MR	Kildwick & Crosshills			North Yorkshire	SE0112345273

Company	Location	Listed	Date	County	NGR
MR	Kimbolton			Cambridgeshire	TL0867271116
MR	Kinnersley			Herefordshire	SO3412648802
MR	Kirkby Stephen		1876	Cumbria	NY7632306482
MR	Kirkstall			West Yorkshire	SE2602335336
MR	Langwathby		1876	Cumbria	NY5742233341
MR	Lazonby		1876	Cumbria	NY5470439894
MR	Liverpool Whitechapel warehouse	II	1874	Merseyside	SJ3462890540
MR	Long Marton	II	1876	Cumbria	NY6676324620
MR	Mansfield Woodhouse		1875	Nottinghamshire	SK5347463243
MR	Matlock		1849	Derbyshire	SK2963360210
MR	Newark warehouse	II	*c* 1875	Nottinghamshire	SK7969854469
MR	Northampton Bridge Street granary	II		Northamptonshire	SP7552759565
MR	Oakham			Rutland	SK8568609077
MR	Oakley	II	1857	Bedfordshire	TL0135953749
MR	Oakworth		1867	West Yorkshire	SE0397538378
MR	Old Dalby		1880	Leicestershire	SK6806723882
MR	Oxenhope		1867	West Yorkshire	SE0326935454
MR	Plumtree		1880	Nottinghamshire	SK6195932347
MR	Raunds			Northamptonshire	TL0206973530
MR	Rushden			Northamptonshire	SP9578667173
MR	Salford Priors		1866	Warwickshire	SP0799051308
MR	Sharnbrook		1857	Bedfordshire	TL0040059554
MR	Studley & Astwood Bank		1868	Warwickshire	SP0594463667
MR	Sutton Park		1879	West Midlands	SP1155596877
MR	Wellingborough	II	1857	Northamptonshire	SP9034468075
MR	Wingfield	II		Derbyshire	SK3851155762
MR	Woodhouse Mill			South Yorkshire	SK4362085801

Company	Location	Listed	Date	County	NGR
MR	Worcester		1868	Worcestershire	SO8587154846
MR	Yate	II	1844	Gloucestershire	ST7011882517
MR/NER Jt	Ackworth		1879	Yorkshire	SE4546017940
MR/NER Jt	Hawes		1878	North Yorkshire	SD8759889869
NBR	Bellingham			Northumberland	NY8416083298
NBR	Chollerton			Northumberland	NY9307871830
NBR	Humshaugh			Northumberland	NY9206570475
NBR	Scotsgap			Northumberland	NZ0388786403
NBR	Wark			Northumberland	NY8712176831
NER	Acklington	II	1847	Northumberland	NU2217301526
NER	Akeld		1887	Northumberland	NT9566929970
NER	Alne		1840	North Yorkshire	SE5044466449
NER	Alne			North Yorkshire	SE5036766586
NER	Alnwick (removed to Beamish)			Co Durham	NZ2160154892
NER	Alston	II	1852	Cumbria	NY7171046771
NER	Appleby		1862	Cumbria	NY6878020807
NER	Askrigg			North Yorkshire	SD9418790851
NER	Aysgarth		1877	North Yorkshire	SE0130489009
NER	Bridlington		1846	East Yorkshire	TA1799066921
NER	Brotton		1875	Cleveland	NZ6859619632
NER	Burton Agnes	II	1851	East Yorkshire	TA1081262414
NER	Carlisle London Road	II	1881	Cumbria	NY4112355008
NER	Cattal			North Yorkshire	SE4468955929
NER	Chester-le-Street		1868	Co Durham	NZ2717351248
NER	Christon Bank	II	1847	Northumberland	NU2133223100
NER	Cloughton		1885	North Yorkshire	TA0114294076
NER	Copmanthorpe		1903	North Yorkshire	SE5672446499

Company	Location	Listed	Date	County	NGR
NER	Cottingham	II	1846	East Yorkshire	TA0508332960
NER	Dacre			North Yorkshire	SE1959061965
NER	Darlington merchandise station	II*	1833	Co Durham	NZ2899215628
NER	Eastgate		1895	Co Durham	NY9574438523
NER	Felling		1876	Tyne & Wear	NZ2729962220
NER	Fencehouses			Co Durham	NZ3178150359
NER	Forge Valley		1882	North Yorkshire	SE9851484530
NER	Forth Bank Goods	II	1906	Tyne & Wear	NZ22455163555
NER	Forth Goods, Newcastle warehouse		1852	Tyne & Wear	NZ2445363781
NER	Fourstones		1838	Northumberland	NY8876667721
NER	Ganton			North Yorkshire	SE9813478593
NER	Gilling		1853	North Yorkshire	SE6155977246
NER	Glanton		1887	Northumberland	NU0818914686
NER	Goathland	II	1865	East Yorkshire	NZ8369801312
NER	Greenhead			Northumberland	NY6593165446
NER	Hedon			East Yorkshire	TA1891929094
NER	Helmsley		1871	North Yorkshire	SE6180283540
NER	Hexham	II	1873	Northumberland	NY9409564247
NER	Hexham Newcastle & Carlisle Railway	II	1835	Northumberland	NY9406064283
NER	Horsforth		1884	West Yorkshire	SE2439139165
NER	Hovingham		1853	North Yorkshire	SE6712576055
NER	Hull Sculcoates		1864	East Yorkshire	TA0971130470
NER	Hunmanby		1853	North Yorkshire	TA1010076667
NER	Hutton Cranswick	II	1846	East Yorkshire	TA0285552265
NER	Kipling Cotes		1865	East Yorkshire	SE9292243922
NER	Kirkby Stephen		1861	Cumbria	NY7703807516
NER	Lartington		1859	Co Durham	NZ0165917805

Company	Location	Listed	Date	County	NGR
NER	Lealholm		1865	North Yorkshire	NZ7613407926
NER	Leeming Bar	II	1848	East Yorkshire	SE2865390015
NER	Levisham lock-up			North Yorkshire	SE8179991026
NER	Leyburn		1856	North Yorkshire	SE1168490284
NER	Loftus		1872–5	Cleveland	NZ7163818051
NER	Marske		1872–5	North Yorkshire	NZ6341821769
NER	Masham		1875	North Yorkshire	SE2325981213
NER	Micklefield		1835	West Yorkshire	SE4446932724
NER	Mindrum		1887	Northumberland	NT8549534009
NER	Morpeth		1879	Northumberland	NZ2042885387
NER	Nafferton	II	1846	East Yorkshire	TA0578258403
NER	Norham			Northumberland	NT9075146758
NER	Piercebridge		1856	North Yorkshire	NZ2121216200
NER	Pocklington	II	1847	East Yorkshire	SE8028448739
NER	Ripon		1871	North Yorkshire	SE3185572314
NER	Robin Hood's Bay		1885	North Yorkshire	NZ9490305413
NER	Saltburn		1872–5	North Yorkshire	NZ6598021330
NER	Sawdon			North Yorkshire	SE9466881738
NER	Scarborough		1845	North Yorkshire	TA0394688277
NER	Selby	II	1834	North Yorkshire	SE6188532284
NER	Shildon		1857	Co Durham	NZ2324725769
NER	Stamford Bridge		1847	East Yorkshire	SE7117955269
NER	Stokesley		1857–8	North Yorkshire	NZ5330107427
NER	Thorp Arch	II	1847	West Yorkshire	SE4382646565
NER	Waskerley			Co Durham	NZ0518945322
NER	Wetherby	II	1847	West Yorkshire	SE4074148704
NER	Whittingham		1887	Northumberland	NU0891812162

Company	Location	Listed	Date	County	NGR
NER	Winston		1856	Co Durham	NZ1404217762
NER	Wistow		1898	North Yorkshire	SE5877735449
NER	Wooler		1887	Northumberland	NT9935328379
NER	York Leeman Road	II	1877	North Yorkshire	SE5935351840
NSR	Ashbourne	II	1852	Derbyshire	SK1762446191
NSR	Blythe Bridge			Staffordshire	SJ9562841163
NSR	Congleton		1849	Staffordshire	SJ8720362344
NSR	Longport			Staffordshire	SJ8551749562
NSR	Rushton			Staffordshire	SJ9361462426
NSR	Sandbach (Ettiley Heath)			Cheshire	SJ7412260296
NSR	Stoke-on-Trent			Staffordshire	SJ8798945503
NSR	Stone			Staffordshire	SJ8965434620
NSR	Tutbury			Derbyshire	SK2137929680
NSR	Waterhouses		1905	Staffordshire	SK0855050162
PD&SWJR	Latchley		1872	Cornwall	SX4018071948
PD&SWJR	Luckett		1872	Cornwall	SX3847771818
S&DJR	Midsomer Norton		1874	Somerset	ST6641053631
S&DJR	Pylle		1862	Somerset	ST6183338871
S&DJR	West Pennard		1862	Somerset	ST5672139601
S&WR	Parkend			Gloucestershire	SO6170907836
SER	Appledore	II	1851	Kent	TQ9756229757
SER	Cranbrook		1893	Kent	TQ7529034565
SER	Dorking Town			Surrey	TQ1594149857
SER	Lydd		1881	Kent	TR0494821524
SER	Paddock Wood		1842	Kent	TQ6698245320
SER	Robertsbridge		1851	Sussex	TQ7341223515
SER	Shalford			Surrey	TQ0029947098

Company	Location	Listed	Date	County	NGR
SER	Canterbury West	II		Kent	TR1461958445
SER	Wateringbury	II	1845	Kent	TQ6907752816
SMJ	Binton		1879	Warwickshire	SP1407653112
SMJ	Ettington		1873	Warwickshire	SP2698450225
SMJ	Helmdon			Northamptonshire	SP5886443845
WSMR	Watchet			Somerset	ST0696443447

Notes

1 Biddle 1997

2 Fawcett 2001–5

3 Fitzgerald 1980

4 Hunter and Thorne 1990

5 Nevell 2010

6 This account is based on Gott 1912

7 Fitzgerald 1980, 30

8 Ibid, 29–49

9 Fawcett 2001–5, **1**, 19–21

10 Wishaw 1842, 237

11 For Andrews, *see* Fawcett 2001–5 and 2011 and Johnson 2013

12 Nairn and Pevsner 1965, 341

13 Tavender 1975

14 Beale 1985

15 Johnson 2013

16 Anon 1913, 29–31; Schloesser and Napper 1901

17 Allison 2014, 114–16

18 Lawrence 1914, 203

19 Lamb 1941

20 Ibid, 79–91

Bibliography

Allison, T 2007 'Industrial building design and economic context: The Chicago Railway freighthouse, 1850–1925'. *Industrial Archaeology Review* **29** No 2, 91–104

Allison, T 2014 'Freight-handling technologies and industrial building design: Freighthouse and warehouse facilities of the Chicago Junction Railway, 1900–30'. *Industrial Archaeology Review* **36** No 2, 109–27

Anderson, V R and Fox, G K 1981 *A Pictorial Record of LMS Architecture*. Oxford: Oxford Publishing Company

Anderson, V R and Fox, G K 1985 *A Pictorial Record of Midland Railway Architecture*. Poole: Oxford Publishing Company

Anderson, V R and Fox, G K 1986 *Stations and Structures of the Settle & Carlisle Railway*. Poole: Oxford Publishing Company

Anon 1913 *Per Rail*. London: Great Central Railway

Archaeo-Environment Ltd 2013 *Statement of Significance for the Former S&DR Goods Shed, North Road, Darlington*. www.aenvironment.co.uk, accessed 5 August 2015

Atkins, T 2007 *GWR Goods Services: Goods Depots and their Operation, Part 2A*. Didcot: Wild Swan Publications

Atkins, T 2009 *GWR Goods Services: Goods Depots and their Operation, Part 2B*. Didcot: Wild Swan Publications

Atkins, T and Hyde, D 2000 *GWR Goods Services: An Introduction*. Didcot: Wild Swan Publications

Beale, G 1985 'The "Standard" Buildings of William Clarke'. *British Railway Journal* **8**, 266–76

Biddle, G 1997 'Goods sheds and warehouses'. *Railway & Canal Historical Society Journal* **32** part 4 No 166, March 1997, 293–9

Bradley, R 2014 'Brick Lane: The London goods station of the Eastern Counties Railway'. *Great Eastern Journal* No 160, October 2014, 4–27

Brees, S C 1847 *Railway Practice, Fourth Series*. London: John Williams

Bywell, E M 1900 'Railway goods depots: No 2 Forth, Newcastle-upon-Tyne NER'. *The Railway Magazine* **6**, February 1900, 149–57

Crofts, C 1998 'Wicker Goods'. *Midland Record* No 9, 46–59

Darley, P 2013 *Camden Goods Station through Time*. Stroud: Amberley

Digby, N L 2014 *The Stations and Structures of the Midland & Great Northern Joint Railway, Volume 1*. Lydney: Lightmoor Press

Digby, N L 2015 *The Stations and Structures of the Midland & Great Northern Joint Railway, Volume 2*. Lydney: Lightmoor Press

Essery, B 1994 'Nottingham Goods'. *Midland Record*, preview issue, 53–60

Essery, B 1996–7 'An introduction to goods stations'. *Midland Record* No 4, 4–24; No 7, 54–75

Essery, B 2003 'Poplar Docks'. *Midland Record* No 18, 7–16

Essery, B 2006 *Freight Train Operation for the Railway Modeller*. Hersham: Ian Allan

Fawcett, B 2001–5 *A History of North Eastern Railway Architecture*, 3 vols (Vol 1 2001; Vol 2 2003; Vol 3 2005). Hull: North Eastern Railway Association

Fawcett, B 2011 *George Townsend Andrews of York: 'The Railway Architect'*. York: Yorkshire Architectural & York Archaeological Society, North Eastern Railway Association

Findlay, G 1889 *The Working and Management of an English Railway*. London: Whittaker & Co

Fitzgerald, R S 1980 *Liverpool Road Station, Manchester: An Historical and Architectural Survey*. Manchester: Manchester University Press

George, A D 1980 'Manchester railway warehouses: A short note'. *Industrial Archaeology Review* **4** No 2, 177–83

Gott, W 1912 'The organisation and duties of the goods department', *in* J Macaulay (ed) *Modern Railway Working, Vol 2*. London: Gresham Publishing, 13–32

Green, J P 1995 'An archaeological study of the 1830 warehouse at Liverpool Road Station, Manchester'. *Industrial Archaeology Review* **17** No 2, 117–28

Grinling, C H 1905 *The Ways of our Railways*. London: Ward Lock

Hunter, M and Thorne, R 1990 *Change at King's Cross*. London: Historical Publications

Johnson, J 2013 *The Survival and Significance of the Railway Goods Sheds of George Townsend Andrews*. MA dissertation, University of York, December 2013

Kay, P 2006 *Essex Railway Heritage*. Wivenhoe: Author

Kay, P 2007 *Essex Railway Heritage Supplement*. Wivenhoe: Author

Kay, P 2012–15 *London's Railway Heritage*, 3 vols (Vol 1 2012; Vol 2 2013; Vol 3 2015). Wivenhoe: Author

Lamb, D R 1941 *Modern Railway Operation*. London: Sir Isaac Pitman & Sons

Lawrence, W F 1914 'Railway goods warehouses'. *The Railway Magazine* **35**, September 1914, 195–203

Medcalf, J 1900 'Railway goods depots: King's Cross GNR'. *The Railway Magazine* **6**, April 1900, 313–20

Minnis, J 1993 'Goods lock ups [of the LB&SCR]'. *The Brighton Circular* **19** No 3, June 1993, 60–4

Nairn, I and Pevsner, N 1965 *The Buildings of England: Sussex*. Harmondsworth: Penguin

Nevell, M 2010 'The archaeology of the rural railway warehouse in north-west England'. *Industrial Archaeology Review* **32** No 2, 103–15

Overton, A E 2003 'Birmingham Central Goods Station'. *Midland Record* No 17, 34–69

Reed, M C 1996 *The London and North Western Railway*. Penryn: Atlantic

Schloesser, H and Napper, W E 1901 'Railway goods depots: Marylebone GCR'. *The Railway Magazine* **8**, January 1901, 54–9

Sheeran, G 1994 Railway Buildings of *West Yorkshire 1812–1920*. Keele: Ryburn

Tavender, L 1975 'Great Western Railway architecture: Goods sheds 1840–1860'. *Model Railways*, May 1975, 222–4

Taylor, S, Cooper, M and Barnwell, P S 2002 *Manchester: The Warehouse Legacy*. London: English Heritage

Timins, D T 1900 'Railway goods depots: No 1 Nine Elms LSWR'. *The Railway Magazine* **6**, January 1900, 70–8

Timins, D T 1900 'Railway goods depots: No 3 Paddington GWR'. *The Railway Magazine* **6**, March 1900, 193–200

Tucker, M 1989 'Bricklayers' Arms Station'. *London's Industrial Archaeology* No 4, 1–23

Vaughan, A 1977 *A Pictorial Record of Great Western Architecture*. Oxford: Oxford Publishing Company

Watling, J 1989 'Bishopsgate and Spitalfields'. *British Railway Journal*, Great Eastern Railway Edition, 85–118

West, F W 1912 *The Railway Goods Station: A Guide to its Control and Operation*. London: E & F N Spon

Wishaw, F 1842 *The Railways of Great Britain and Ireland*. London: John Weale. Reprinted 1969, Newton Abbot: David & Charles

Informed Conservation series

This popular Historic England series highlights the special character of some of our most important historic areas and the development pressures they are facing. There are over 30 titles in the series, some of which look at whole towns such as Bridport, Coventry and Margate or distinctive urban districts such as the Jewellery Quarter in Birmingham and Ancoats in Manchester, while others focus on particular building types in a particular place. A few are national in scope, focusing, for example, on English school buildings and garden cities in England.

The books are written in an engaging style and include high-quality colour photographs and specially commissioned graphics. The purpose of the titles in the series is to raise awareness in a non-specialist audience of the interest and importance of aspects of the built heritage of towns and cities undergoing rapid change or facing large-scale regeneration. A particular feature of each book is a final chapter that focuses on conservation issues, identifying good examples of the reuse of historic buildings and highlighting those assets or areas for which significant challenges remain.

As accessible distillations of more in-depth research, they also provide a useful resource for heritage professionals, tackling, as many of the books do, places and building types that have not previously been subjected to investigation from the historic environment perspective. As well as providing a lively and informed discussion of each subject, the books also act as advocacy documents for Historic England and its partners in promoting the historic environment through the management of change.

More information on each of the books in the series and on forthcoming titles, together with links to enable them to be ordered or downloaded, is available on the Historic England website: HistoricEngland.org.uk